The **Eco-Home**
Design Guide

Praise for The Eco-Home Design Guide:

"In *The Eco-Home Design Guide*, Christopher Day makes the technical simple even for non-scientific readers. With gentle humour, this most gifted of architects shares his vision of a living architecture, demonstrating that eco-design CAN be supremely nourishing for the soul." **– Susan Swire, Author of *Secrets of the Feel-Good Home***

"You could hardly find a better man to lead you through the challenging process of creating an eco home from scratch, or of renovating an existing building to meet eco criteria." **– Angela Neustatter, journalist and author of *A Home for the Heart***

"Christopher Day is one of Britain's most outstanding holistic and ecological architects and this book is a timely gift when freeing ourselves from the dependence on fossil fuel has become an urgent imperative."**– Satish Kumar, Editor-in-Chief, *Resurgence & Ecologist* magazine**

"This book is an ideal guide for those who wish to explore to the greatest depth the possibilities for designing their own eco-homes." **– Professor Chris Tweed, Welsh School of Architecture**

"Christopher is the UK's foremost practitioner of soulful architecture and place-making. I would urge anyone considering building a new house to read his latest gem, The Eco-Home Design Guide." **– Peter Stirling, Managing Director, Stirling Developments Ltd**

"The generous use of illustrations and side notes makes this an easy to use guide, and it is a must-have reference for every architect, homeowner, and developer in the operation and construction of homes, retail, and office buildings." **– Richard Erganian, developer**

"A delightfully accessible reach-for book. I find so many people blundering and wandering in circles on their eco house creation journey, this is just the thing they were missing." **– Vicky Moller, Welsh Assembly election candidate for Plaid Cymru, Party of Wales**

"A unique text in the study of the elements which constitutes effective, sustainable architectural design for healthy living." **– Lino Bianco, Senior Lecturer at the Faculty for the Built Environment, University of Malta**

"Besides widening the eco-design debate beyond simplistic energy-conservation, this book describes complicated issues clearly enough for the most non-technical reader to understand." **– Prof. Dr Justyn Ochocki, Medical University of Lodz**

The **Eco-Home** Design Guide

Principles and practice for new-build and retrofit

CHRISTOPHER DAY

with a foreword by HRH The Prince of Wales

Published by

Green Books
An imprint of UIT Cambridge Ltd
www.greenbooks.co.uk

PO Box 145, Cambridge CB4 1GQ, England
+44 (0)1223 302 041

First published in 2016, in England

Christopher Day has asserted his moral rights under the
Copyright, Designs and Patents Act 1988.

Interior illustrations by the author and Jindarah Chatuchinda
Cover illustration by David Mostyn

Design by Jayne Jones

ISBN: 978 0 85784 304 3 (hardback)
ISBN: 978 0 85784 305 0 (paperback)
ISBN: 978 0 85784 306 7 (ePub)
ISBN: 978 0 85784 307 4 (pdf)
Also available for Kindle.

Disclaimer: the advice herein is believed to be correct at the time of printing, but
the author and publisher accept no liability for actions inspired by this book.

10 9 8 7 6 5 4 3 2 1

To Olusia

Acknowledgements

I'm deeply indebted to the many people I have worked with and from whom I have learnt much – sometimes by making mistakes *I* won't repeat but *they* must live with. In particular, I must thank my editor, Alison Kuznets, for her perceptive observations and suggestions, Joachim Eble for teaching me most of what I know about healthy building, Sue Roaf for her support and advice over many years, Bill Holdsworth for stimulating thoughts about cooling, and Richard Erganian for his insights and access to his extensive library. Also, Vicky Moller for reading my draft manuscript and making pertinent suggestions, Martyna Kieliszek for finding bits which needed clarification and Dewi Day for reviewing technical information. Finally, especially, my wife, first for identifying the need for a book like this and then inspiring me to write it; and, doubly especially, for her help and support, without which I could not have done so.

Contents

Preface

This book is written on the basis of almost forty-five years' commitment to, and practice of, eco-architecture (including designing two urban eco-developments, five eco-villages / neighbourhoods and over a hundred and fifty homes); over thirty years of living and working in different climates (extremely cold, extremely hot, extremely windy and wet); nearly thirty years of hands-on self-building; twenty-five researching and writing about eco-design (although in the first fifteen, there was little to read); and over fifteen years living with, and adapting buildings for, disability. I've had plenty of opportunities to make mistakes – from which I've learnt a lot – and I hope to live long enough to make more, and so to learn more. The biggest mistake, of course, is to regard the relationship between humanity and nature as a battle. Fortunately, I've been spared from this by spending much of my life building in inclement weather. (Against weather, you'll always lose a battle; it's more productive to work out how to live with it.)

Additionally, I've improvised, made and used many things that (at the time) I hadn't read and learnt about; and (since then) have read and learnt about some things I haven't used or handled. (Handling is more important than using, as you learn more through your hands.) Consequently, neither I, nor my publishers, can accept responsibility for any errors, omissions or misinformation.

As I am writing in the northern hemisphere, I use south to mean towards solar zenith, and north, away from it. Readers in the southern hemisphere should reverse this, preferably by reading the book upside down.

Note on quantities

These are first given in metric or imperial according to their source or most common usage. Likewise, when converting between units, figures have been rounded to match their source, if rounded. As a handy tip, conversion between imperial and metric is simple using the 10% rule:

1 metre = 1 yard plus 10%
1kg = 2lbs plus 10%
1 litre = 2 (imperial) pints minus 10%
An increase of 1°C = an increase of 2°F minus 10% (but water freezes at 0°C or 32°F, depending where it is).

Similarly, as there's always more of everything in America, there are more US gallons: they're around 20% smaller than UK gallons. US and UK billions also differ, but the common factor is 'lots'. For gallons, it's 'wet'.

Brief biography

Born: 1942. Nationality: Welsh.

Studied architecture, then sculpture in London. Committed to eco-architecture (and ecological lifestyle) from 1970. Built four eco-homes (by hand: no machinery – never again, please!), then led volunteers to build two Steiner schools.

Became a (self-employed) architect by a disreputably unconventional, indeed unprecedented, route (designing and building, not passing exams). Received four design awards (including a Prince of Wales Award). Pioneered a consensus design method – which I have used for over sixty projects – based on *listening*: to people (often opinionated), place and situation (each one, unique). Visiting professor, Queen's University, Belfast, 1993–8.

Have worked (design, consultancy, lecturing, teaching) in 21 countries (25 if you count Texas, Québec, Kernow and Siberia as independent, but 20 if Crimea is part of Russia), a wide range of climates, cultures, wealth and expectations.

Eight other published books (and German, Italian, Russian, Thai, Czech, Greek, Spanish and Welsh translations).

I have long believed that our built environment has an enormous impact on the quality of our lives and our well-being, as well as on the nature of our communities. As our planet becomes overwhelmingly urban and natural resources become scarcer, it is clear to me that if we are to build new communities to improve the lives of the people who live there, then significant emphasis must be placed on the design of these homes and their use of natural resources. It is vital that we construct our homes not just as a haven from the hustle and bustle of life, but as protection from increasingly erratic elements, making them as efficient (and inexpensive) as possible to heat or cool and with an eye for beauty and attraction that stands the test of time.

It is for this reason that I am so delighted to have an opportunity to contribute to Christopher Day's handbook, 'The Eco-Home Design Guide', having long admired his previous publications, including 'Places of the Soul', which I thought extremely informative. This book is no exception and is exceptionally timely.

Although experts have told us for years that homeowners can add various gadgets to make homes more eco-friendly, putting turbines here and perhaps solar panels there – what might be called an 'eco-bling' approach – all the evidence I have seen is that it simply doesn't solve the fundamental problem, which is the need to build simple, beautiful, efficient homes. Rather than such short-term, technologically driven approaches, why not build homes genuinely designed to demand little or no energy from the outset? I am so glad that Christopher Day's book addresses these issues.

Some years ago my Foundation for Building Community, together with the Building Research Establishment, designed the Natural House in order to demonstrate the most effective route to low-energy, low-carbon homes designed for longevity and with traditional appeal. However, as Christopher's book explores, this remains an area of great debate. What is an eco-house? How should one measure performance and should an eco-house look like a space-ship or something that can exist harmoniously in a traditional village? An equally complex debate is taking place about making existing homes more energy efficient. Should you seal up the windows (God forbid!)? How do you make an old building airtight? What is the difference between airtightness and a breathable building?

This is clearly an area where a careful understanding of the biology of the home, together with consideration of what a home is for – namely housing people – is vital. All too often, the performance of a home is all about the measurement of temperature or energy, when in fact these might have negative impacts on the way people feel or want to live. So rather than thinking about temperature, maybe it is better to think about comfort; rather than thinking about air changes, perhaps an analysis of air quality and health is important, and so on. All of these factors should be analysed and discussed with people at the heart of the design process.

It seems to me that when we work with Nature, rather than in opposition to her, we can find inventive solutions that blend the best of ancient knowledge with modern needs. Nature offers us a rich palette of attractive, efficient materials as well as potential gains from the power of the sun, rain and wind. I believe we tend to overcomplicate the construction of environmentally responsible buildings, when we should strive to find simple, natural and attractive solutions.

It is therefore a delight to see this eco-home design handbook tackle so many of these critical issues in a way that makes them both human and natural, providing a practical framework for much that is dear to people and of vital importance for our children's and grandchildren's future. I hope everyone using this guide will take heed of the extensive advice and that it will direct our thinking about living more thoughtfully and sustainably on the one planet we have.

CHAPTER ONE

Introduction: why, where, how and when to use this book

WHERE? (Microclimatic, social, connective and site planning issues) page 20

HOW? (Keeping warm, cool, dry, healthy and safe) page 46

WHAT? (Minimising hidden environmental impacts, and generating energy) page 192

WHEN AND WHO? (How to, and who should, build; and in what sequence?) page 218

WHY? (What benefits do you seek?) page 16

What is an eco-home? What do you need to know to create one?

What is an eco-home? If you more or less know, and more or less know how to create one, you don't need to read this book. Or perhaps you do. There may be things you haven't thought about, or aspects of things that you haven't considered in depth or didn't know. Some might even be vital issues for how your house performs and lasts, and how it is to live in – and how long *you* live.

Everybody more or less knows what an eco-home is, but as there's no official definition, nobody *actually* knows. Is it necessarily a house? Can it be a cara-van[†], yurt or even a high-rise apartment? Is it a zero-carbon[†] – or, better, a carbon-negative[†] – house? *(Glossary entries are indicated[†] at their first occurrence.)* Would that include a wholly uninsulated house connected to a photovoltaic[†] farm? Or is it a house so nature-friendly that it's full of wasps, rats and (suspiciously ominous) mushrooms? Or should it be so zero-impact that living in it means dressing for outdoor temperatures? Everybody knows it doesn't mean those sorts of things.

It's safer, therefore, to focus on what everybody *more or less* knows: that it's something to do with supporting ecological stability.

What is ecological stability, though? Climate stability is an absolutely essential part of this, but only a part. Preserving or recycling key resources (e.g. water, nutrients, perhaps manufacturing feedstocks) is comparably important, although this hasn't yet reached a tipping-point crisis. So is the maintenance of a self-regulating ecology. This is mostly about reducing pollution, preserving biodiversity[†] and, indirectly but crucially, food security. Common to all these, however, is living in harmony with nature. But can we do this if we live in town? As most of us do, for life to survive on our planet, this is absolutely essential: living *in harmony with* nature doesn't mean living *in* nature.

Historically, however, the eco-architecture movement didn't start very harmoniously. There were those who focused on minimising energy use. Hermetically sealed buildings with tiny windows appealed to them. Others were principally concerned with occupant health. They liked air-permeable buildings, lots of fresh air and solely natural materials. Others again sought hi-tech solutions to everything. Some focused on autonomous houses and self-sufficient lifestyles; others, on social issues – even to the point of collectivism. And everybody thought 'ecology' meant they couldn't use their favourite materials or do anything artistic. Nobody ever considered the risk of dying from heatstroke, at least not in Britain: at that time, heat retention (for resource preservation) seemed the issue, not cooling (for survival). Climate change wasn't a concern; oil depletion (and related price rise) was. Also, although some thought meditation could transcend ecological disaster as this is 'merely' material, nobody discussed the spiritual dimension of sustainability.

Fortunately, those days are behind us – but their echoes still remain. Few seem to realise that harmony with nature means embracing and synthesising *all* of this. If we omit the material concerns, things don't work. If we omit the aesthetic, we only serve the material side of nature and humanity. The climate crisis demonstrates this. If material solutions don't work, human – and probably all – life won't last. But if concentration on practical aspects ignores soul and spirit issues, life won't be worth living – and (according to statistics) is likely to be shorter.[1]

Does this make what "everybody more or less knows" unduly complicated? It shouldn't. It just means that there are many more issues than energy conservation to attend to. In this book, I therefore first address the 'why' behind eco-home design: the overarching 'mother issue'. Then follow 'where', 'how' and 'what' issues: the specifics. Further, to actually build or convert your home I also discuss the 'when' and 'who' issues: achieving its delivery.

How to find what you need in this book

This book is divided into six parts. These cover the reasons – and their design implications – for wanting to live in an eco-home; how to decide where it should be and remedy the shortcomings of your chosen location; how, in performance and constructional terms, to design a new eco-home or eco-upgrade an existing building; the things it affects and the energy it needs; how to do it; and how to get started and avoid mistakes.

WHY ISSUES helps you to identify the particular benefits you're aiming for. These are much more varied than most people, looking through their individualised lenses, suppose. They range from monetary to spiritual, personal to global. We may want them all, but different people have different priorities. This affects design.

WHERE ISSUES covers microclimatic, social, connective and space-use matters, and how these affect site planning. These also have much greater significance and scope for action than is generally supposed. 'Where' also applies to eco-renovation, but in a different way: you need to be able to evaluate problems that existing buildings have. All old buildings have some, but how easy are they to remedy? This is critical to your decision about what building to buy and eco-convert.

HOW ISSUES introduces you to how to get your home to do all those things you need it to: keep you warm, dry, cool, healthy and safe. Whether you're building an entirely new eco-home or renovating a less-than-satisfactory old building, the basic physical and design principles are the same. Consequently, although chapters on old buildings deal with specific renovation issues and associated constraints, pitfalls and solutions, the bulk of every chapter is relevant for both. As there's a huge variety of building types, construction and ages, and of climatic circumstances, this is the largest section of the book.

WHAT ISSUES covers hidden environmental impacts; recycling of water, waste and nutrients; and generating energy.

WHEN AND WHO ISSUES discusses how to proceed, to actually build or convert your home. Who will build it: you or a builder? What are the practical and sequence-related implications of your involvement? The final chapter helps you get going!

Before you start, I recommend you read the **Case studies** (Appendix 1, page 232). These give examples of eco-homes I have designed and what I've learnt from the mistakes I made. I suggest you read the descriptions and test yourself to see if you can anticipate the mistakes described in the **What I should have done** reviews.

Most chapters includes a list of resources (in alphabetical, *not* importance, order). These include further information on sustainable building issues, techniques and products. This, of course, can only be a partial list. For the products especially, there is a huge amount of information out there. For example, a complete list of all the (purportedly) ecological building products found at Ecobuild (the largest building trade exhibition in Britain) would fill several volumes. (Instead of this, just visit *www.ecobuildproductsearch.co.uk*) Moreover, new products, suppliers and publications are appearing all the time, so this list is unavoidably already incomplete and out of date. Consequently, I have limited the products and suppliers listed here to those less common ones that I have used or considered using. Many items are relevant across different chapters but, unless of particular importance to another chapter, aren't repeated.

Unavoidably, this book contains some technical terms and concepts you may be unfamiliar with. I have included a glossary at the back of the book, so you can both understand these terms and have sufficient techno-speak to convince builders you understand enough about building so they can't cheat you. (You don't have to understand, and they may try to cheat – but it still helps.)

Speed-reading

The time for abstract theorising disappeared with the last century. Climate change is now an established fact and one that we have to find out how to live – indeed, to survive – with. We no longer have time to spare, so we need to know the practical ways of doing things. This book aims to help get you started. To speed understanding and highlight crucial practical issues, I include choice-tables, keypoint summaries and a large number of diagrams and illustrative examples. To further assist speed-reading, and to help navigate your way through the book, each part and each chapter starts with a diagram of the issues it covers.

To create an eco-home, it isn't necessary to read the whole book. Just look at the sections you need to know more about. However, to create an eco-home that's *worth living in*, I recommend you *also* look at the sections you think you *don't* need. The topic, after all, is immense. Moreover, as every family is unique, every home (and every eco-home) will be unique. No one formula can ever fit all. Only you know your needs and preferences, and the unique location of your home gives unique opportunities. That's why you need to start, not with fully formed answers, but by asking the right questions. This book, therefore, aims to give a brief and holistic overview, so that you can ask the questions right for your situation. The next step is what to do, then how to do it. I therefore include enough practical details and concrete examples so that you can find out most, perhaps all, of what you need to know, look in the right direction – and see through salesmen's hype – if you want to know more.

---

1. By 15 years on average: Jüngel, S. (2010) Culturally active people live longer, *Anthroposophy Worldwide*, No. 5/10.

I. WHY ISSUES

Minimising ecological damage (page 17)

Minimising climate damage (page 17)

Soul-nourishment (page 17)

Resilience to weather extremes (page 17)

Health (page 18)

Resilience to energy-supply disruptions (page 17)

Lifetime use (page 18)

Thermal comfort (page 17)
Energy bill reduction (page 18)

Why have an eco-home?

CHAPTER TWO

Overarching reasons

Why build an eco-house or eco-convert an old house? Is it for your benefit? Your children's and grandchildren's? The world's? Or all of these?

Most of us feel a measure of eco-responsibility to do something to mitigate climate change. The built environment accounts for about half of all climate and ecological damage, and housing accounts for 30% of all human-caused CO_2 produced.[1] The design and construction of our homes and how we live in them, therefore, give considerable scope for reducing this. Mostly, reducing climate damage is about reducing heat loss. In Europe, 75% of building energy heats homes,[2] and in Britain 80%.[3] Energy-efficient lights and appliances – and energy-minimising use of them – can significantly reduce energy consumption, and also bills! Furthermore, as summers become hotter, cooling will require increasing amounts of energy. Air-conditioning is electrically powered. Most electricity requires three times more fuel per kW than heating.

Embodied energy and pollution in the materials buildings are made of are also significant. Nowadays buildings embody about a third of the energy that they use over their lifetime. European buildings typically last a hundred years,[4] American ones generally much less; but the longer any building's lifetime, the longer period its embodied energy is amortised over. Longevity doesn't only depend on durable materials; adaptability to changing needs (e.g. family size, homework, other uses) and easy reparability are also factors. Most important, however, is that the building is sufficiently soul-nourishing to be *worth* repairing. Also important to remember is that sustainability is about much more than energy: it's about "living on the planet as if we intend to stay".[6] This has social, ecological and aesthetic dimensions.

What benefits do you seek?

Sustainability is often the principal reason for wanting an eco-home, but it's rarely the sole reason. Sustainability is about benefits to *others*: our children, grandchildren and great-grandchildren. But it's *we* who pay the considerable sum that building, buying or renovating *any* house involves. So, as we are the ones who pay, what's in it for us?

There are many benefits from living in an eco-home. Eco-homes are more comfortable thermally, cheaper to keep warm or cool and more resilient to weather extremes and energy supply disruptions. They're also healthier to live in than 'normal' homes. Your primary motivation for creating an eco-home may be any one of such benefits.

Many people's main objective for their eco-home is thermal comfort: somewhere cosily warm in winter and deliciously cool in summer. With climate change already increasing weather extremes in many places, this becomes increasingly important. We shouldn't assume that global warming means warmer winters. It *might*; but the regional effects of global temperature rise will be complex. For instance, if the Gulf Stream – already weakening – changes course, winters in north-west Europe will be *much* colder. Nor should we assume that keeping heat *in* is the only issue: summer heatwaves are already becoming more common across the world. Over 30°C (86°F), these start to lose their appeal; over 45°C (113°F), we never want to experience another one. This is about more than comfort: for the vulnerable, it's a survival issue. Related to this is future-proofing: building habitability when energy supply – particularly electricity – fails. Most heating and cooling systems depend on electricity (e.g. for pumps and fans). Extreme weather, oil depletion and international politics, however, mean uninterrupted electricity supply is no longer assured.

Another common objective for creating an eco-home is to save money. With oil depletion raising energy prices, energy bill reduction becomes increasingly attractive. Eco-buildings are dramatically cheaper to run. They're usually more expensive to build – sometimes a little more expensive, sometimes a lot more – but not always; some are no more expensive, or even cheaper than conventional buildings.[6] In lifetime-cost terms, however, reduced energy bills outweigh any increased construction expense many times over. If we design for disabled access, we may well be able to live in our house for a lifetime. And if we need to move before recouping our costs, resale value increasingly reflects a house's low-energy benefits.

It's possible, however, that excessive focus on energy conservation can produce environments that are unhealthy to live in – what William McDonough calls "killing machines".[7] Since Sick Building Syndrome (SBS) began to be recognised in the 1980s, there has been growing awareness of the effect buildings can have on health. Many building (and furnishing and cleaning) materials emit small amounts of toxic gasses. In the past, draughts and uncontrolled ventilation diluted these but also lost a lot of expensive heat. The more airtight buildings are, the more important indoor air quality becomes. Additionally, modern building materials contain many more unstable chemicals than traditional ones do, particularly glues and plastics. Occupant health is a growing issue, especially for children, whose immature organs are more vulnerable to these toxic compounds. This isn't a solely modern problem: airborne chemicals may be modern, but many old buildings – and even some new ones – are damp, and therefore full of mould spores, which aren't good to breathe. Additionally, human sickness and health are complicated. Besides physical nutrition (air, water, food and warmth), health is influenced by social, sensory and psychological factors. Healthy building must consider all these.

Most of us want to live longer, and remain healthy. We therefore need healthy homes. This has both passive and active dimensions: from non-toxicity and thermal comfort to well-being induced by social, nature connection and multi-sensory aesthetic factors. We might consider the fitness advantages of being able to walk or cycle to work, or even of maintenance-intensive gardens or three-storey homes (although stairs are a major cause of accidents[8]). The longer we live, however, the more likely we are to experience mobility, sensory or memory disabilities. Consequently, we'll need homes we can move around in and manage when we're infirm. (We could, of course, just move to somewhere better, but moving home is the third most stressful event in life,[19] and old age the least healthy time to be stressed.) This means homes should ideally be fit for – or be easily adapted for – lifetime use, namely fully disabled-accessible.

Eco-purism or pragmatism?

Most of us are concerned about all these issues but to varying extents. Whichever we prioritise will affect how we approach eco-house design. It's unwise, though, to ignore any of them completely. It's certainly easy to build a house that uses no energy at all – a sealed polystyrene box will do – but no home is worth living in if it's miserably uncomfortable, depressing or will kill us – or will help kill the planet.

Eco-purism can fulfil many aims. It is possible to reduce CO_2 emissions by over 90% and other impacts even more. (Indeed, a wood-fire-warmed, hide-draped tipi causes no net CO_2 emissions at all to build or live in.) Such impact reduction is inspiring, and it feels good to live in an eco-purist home. However, few of us want to over-winter in a tipi, or can afford to buy a purpose-designed, eco-purist house. Nor can all of us afford to do everything. Even in an old house, an 80% reduction of environmental impacts is reasonably achievable and affordable, but is this aim a compromise? Eco-purist inspiration is indispensable to motivate change, but ten (affordable) 80% achievements deliver eight times as much as one (expensive) 100% achievement. *Both*, therefore, are essential.

Whatever the balance you choose between eco-purism and pragmatic affordability, you'll make a significant contribution to a more sustainable world – and reap personal benefits too.

Keypoints

- Lifetime costs dwarf first costs. Assess cost/benefit on this basis.
- Consider thermal comfort in the light of climate change, rising energy prices, and energy supply disruptions. Recognise it as a potential survival issue.
- Consider your home's effect on health, accessibility (if age brings infirmity) and on our planetary responsibilities.

Choices

Aim	Focus
(Lifetime) cost-saving	Energy bill reduction of 80% or more.
Comfort	Optimal thermal performance.
Future-proofing	Minimal energy/electricity dependence.
Inspiration	Eco-purism, soul-nourishment, beauty.
Climate protection	Carbon-negative design and operation.
Well-being	Healthy building; sensory, soul and spirit nourishment; greenery; privacy but community.
Environmental impact reduction	All of the above; also minimal embodied-energy / pollution materials, cycle-closure, appropriateness to context.

Resources

Overview and application to all or most issues

AECB: *Green Building* magazine: well worth subscribing to this journal.

Broome, J. (2007) *The Green Self-build Book*, Green Books, Devon.

Day, C. (3rd ed. 2014) *Places of the Soul*, Routledge, London

Day, C. (2002) *Spirit & Place*, Architectural Press, Oxford

Hall, K. (ed.) (2008) *Green Building Bible*, Vols. 1 & 2, Green Building Press, Llandysul.

Roaf, S. (2001, 4th ed. 2013) *Eco-house: A Design Guide*, Architectural Press, Oxford.

www.carbontrust.com (for larger-scale projects only)
www.ecobuildproductsearch.co.uk
www.energysavingtrust.org.uk
www.greenspec.co.uk

1. In UK: Cotterell, J. and Dadeby, A. (2012) *PassivHaus Handbook*, Green Books, Cambridge.
2. VDEW (2002) Final energy consumption 2002; electricity, oil, gas, coal, etc. In Britain, of all buildings' energy, heating alone takes 47%: *Today* programme, BBC Radio 4, 2 February 2009. Homes use 50%: Energy Saving Trust.
3. Hall, K. (ed.) (2008) *Green Building Bible*, Vol. 2, Green Building Press, Llandysul.
4. Glücklich, D. Neuhäuser, M. (2000) Measurable sustainability with the 'building passport'?, in Sala, M. and Bairstow, A. (eds) Roaf, S. TIA Conference 2000 proceedings; Svane, Ö. (2000) Trapping the impacts of existing buildings, in Roaf et al. TIA Conference 2000 proceedings.; Vale, B. and R. (2009) Sustainability begins at home, *Building Design*, 20 November 2009; Cotterell, J. and Dadeby, A. (2012) Now, however, they're built for sixty, in *PassivHaus*.
5. Sir Crispin Tickell.
6. Once past unfamiliarity (for which some contractors double price), additional costs for PassivHäuser decline to 3.7% or even zero: News Briefings: First certified non-domestic PassivHaus in UK, *Green Building* magazine, Autumn 2009, Vol. 19, No. 2; de Selincourt, K. (2013) The cost of building passive, *PassiveHouse+*, 2013, Issue 3. Even in 2005, BRE estimates of the additional costs of sustainable building were only 1.1-1.8% of typical house prices: Pembrokeshire County Council (2005) Supplementary planning guidance, building in a sustainable way. 'Normal buildings', however, mean those that meet current energy conservation standards, not 'common buildings' – which fall woefully short.
7. Lazell, M. (2009) Architects are creating toxic 'killing machines', *Building Design*, 27 March 2009.
8. *http://www.rospa.com/homesafety/adviceandinformation/general/facts-figures.aspx* (accessed: 23 October 2014).
9. Employee Relocation Council, cited in *3rdculturechildren.com/2013/04/08/packing* (accessed: 9 April 2014).

II. WHERE ISSUES

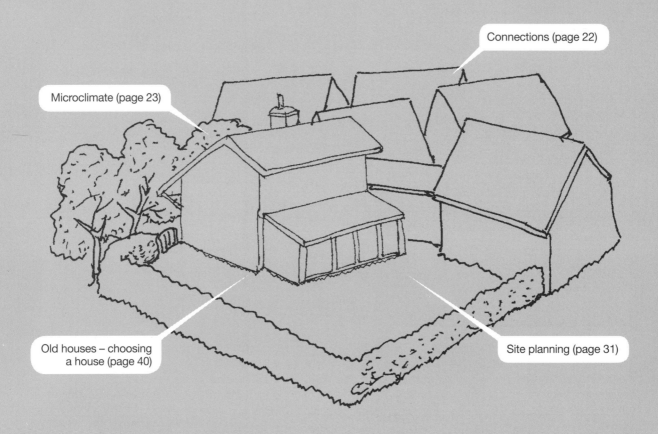

Connections (page 22)

Microclimate (page 23)

Old houses – choosing a house (page 40)

Site planning (page 31)

Before you can build an eco-home, you need somewhere to put it. Similarly, before you can eco-convert a building, you need to find one. For both, this requires choosing a suitable location, but what does 'suitable' mean? Of the many things influencing this, climate (or, more precisely, micro-climate[†] – which we can do lots about) significantly affects our well-being and energy consumption, hence ecological impact.

Although this section is largely focused on new buildings, many of the issues are equally relevant to old buildings. Issues that are specific to an already-built building are covered in Chapter 3 (page 21).

Climatic and microclimatic considerations

CHAPTER THREE

Wind (page 21)

Trapped-air insulation (page 24)

Sun (page 24)

Shade (page 23)

Local climate basics

As winds transport heat, cold and moisture, winds from different directions are typically warm or cold, dry or damp – modified, of course, by season. In Britain, north and east winds (from the Arctic and northern Siberia) are often cold and usually dry, whereas south and west winds blow across the Gulf Stream-warmed Atlantic, so are usually warm and often wet. These are the more common 'prevailing winds'. Consequently, as major hills lift air to cooler levels where water vapour condenses, their south-west slopes are usually cloudier and get more rain than their north-east sides.

This applies to high-level winds. At ground level, however, hills, woodland and buildings shelter from, deflect or funnel winds. This causes winds to blow along, not across, valleys and streets. In windy areas, long views – especially sea views – mean increased wind exposure. Windbreaks (whether topography, trees, hedges or buildings) reduce wind speed. However, just like gaps under doors, gaps in those windbreaks increase wind speed, as do hilltops and ridges – all with a corresponding increase in wind-chill.

Orientation and air temperature (for the British Isles and much of Europe)

- East is almost as cold as north – sometimes colder as east winds are more common.

- West is almost as warm as south – sometimes warmer as, in summer, air temperatures peak mid-afternoon.

- South, however, has longest sun exposure. North has hardly any – and only in summer.

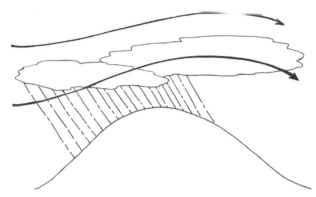

Hills' south-west slopes get more rain than their 'rain shadows' to the north-east.

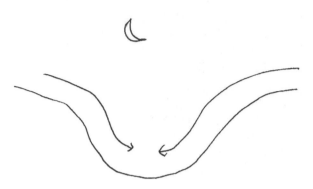

As everything radiates heat to space on clear nights and cold air is heavier than warm, it drains down slopes. Valley bottoms can be 10°C (18°F) colder at night than their sides.

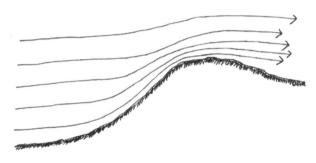

Hilltops can increase wind speed up to 70%,[1] or even double it.[2] Only those with windward sides steeper than 1:3, however, produce turbulence to lee.[3]

South-facing slopes get stronger and longer sunshine than north-facing ones.

Conversely, valley bottoms are sheltered from most winds; this is advantageous in winter but sometimes stifling in summer. Valleys aren't always the warmest places, though. As cold air is heavier than warm, it drains down slopes, even shallow ones. So, valley bottoms can be 10°C (18°F) colder than valley sides on still cloudless nights.[4] In winter, these cold air ponds can remain as 'frost pockets' throughout the day, especially if shaded. Also, north-facing slopes and steep-sided valleys get much less sun. South-facing slopes get more intense and longer periods of sunshine than north-facing ones. This makes mid-slope south-facing valley sides the warmest places to live, providing they're sheltered from wind.

These factors, along with thermal-buffering vegetation and humidity-retaining woodland and marsh, affect a broad area's mesoclimate[†] (e.g. town, vale or coastal-strip climate) and, more locally, a specific site's microclimate. Microclimate affects plant growth, animal behaviour, our use and enjoyment of places – and heating/cooling bills.

Nonetheless, for our well-being, our choice of area is normally guided by factors like proximity to friends, work and schools. Living close to these amenities reduces travel. This usually achieves a greater CO_2 (and monetary) reduction than choice of home location by site-climatic factors alone. Proximity is often limited by availability of property within our price range. Microclimate, however, we can alter. For ecological design, it's the foundation on which all site-selection, layout, design and landscaping decisions need to be built.

Keypoints

- Microclimate is possible to alter.
- Topography, woodland and windbreaks shelter from wind.
- Topography, woodland-edges and streets deflect winds.
- Long views increase wind exposure.
- On clear nights, cool air drains down slopes and ponds on valley floors.

Choosing what should be where in the garden: microclimatic landscaping

However hot, cold or savage the weather, microclimatic landscaping can moderate it. How, where and with what you plant your garden can alter its microclimate and hence the temperatures around your house. This can both increase comfort outdoors and reduce the need for heating and cooling indoors.

Different types of tree can serve different microclimatic purposes. Individual trees can break the force of wind; and tree rows and hedges can give shelter from it. Tree windbreaks reduce wind speed, therefore reducing both destructive capacity and wind-chill; such windbreaks can lower heating bills by up to 30%.[5] But what if wind exposure coincides with view? For this, you can cut 'windows' through trees and deflect wind with shrubs. Such windows, of course, need repeated (sometimes annual) pruning as foliage tends to fill in gaps.

Narrow passages between houses and outbuildings *accelerate* wind (as door frame gaps do), increasing wind-chill. To minimise heat loss from buildings, therefore, such narrow passages need wind-screen-covered or slatted gates (optimally 60% solid, 40% gaps). Also, as air currents are rather more pushed (which is directional and has momentum) than vacuum-pulled (which is less directional),[6] plant a draught-stopping bush at the passage's windward end. Alternatively, you can use such passages to dry laundry. Similarly, the wind down them can drag air out of windows in the buildings lining them for cooling in summertime. Gates allow you to choose draught or (relative) calm according to season and need. Breeze acceleration by tree-blocks, bushes, walls or buildings can also be used to scour snow from doorways in winter or to direct cooling airflow into the house in summer.

In hot weather, trees give cooler shade than awnings. Unlike fabrics, leaves don't get hot, and they cool by transpiration. Unlike solid shades, they allow some air movement. As solar heating is only needed in winter,

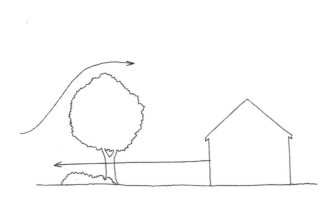

'Windows' cut through wind breaking trees.

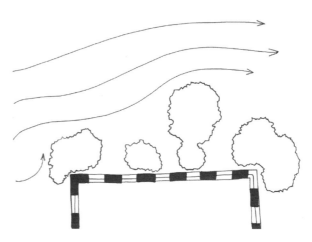

Breeze deflected by shrubs.

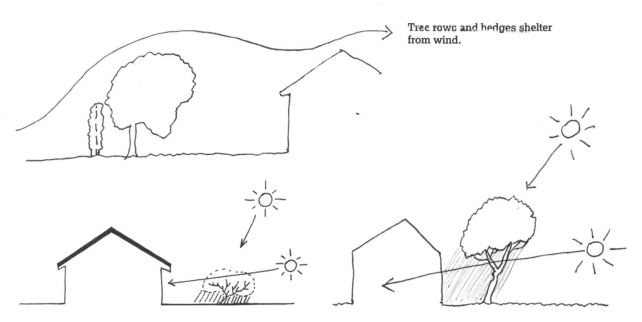

Tree rows and hedges shelter from wind.

Low-spreading trees shade in summer but hardly in winter.

High-crowned trees with clear-pruned trunks shade houses in summer but admit winter sunlight.

trees for shading south windows in summer can be placed and shaped to admit low-angle sun for warmth indoors in winter. Low-spreading trees (e.g. apple) or high-crowned ones with clear-pruned trunks shade in summer when the sun is high, but in winter cast little shade where it matters. Fruit trees are pruned annually for light admission; other sorts should be light winter-twigged.

Theoretically, your south boundary should be open to sunlight but north boundary protected from cold wind by windbreak trees. But not only does weather defy such simplistic categorisation, it's also important to remember that every north boundary is someone else's south boundary. Is optimum microclimate worth bad neighbour relations? Indeed, this principle applies widely: everything you do to the land outside your house, and also to the outside of your house, affects your neighbours – their microclimate, views and mood-of-place. Ultimately, we have only one world to share.

It's not only wind and wind-borne air that cools us. The ground radiates heat to dark space at night, irrespective of air temperature. That's why there can

be ground frost without air frost. Clouds obstruct this night-time cooling, keeping warmth in. Leaf canopy overhead has a similar effect, keeping ground, air, and therefore buildings, warmer at night. Evergreen species, like pine, holly, and holm oak, obstruct this radiant cooling better than dense-twigged species. Cold ground cools the air immediately above it. This cold air drains downhill but you can choose where it goes. Solid obstructions (e.g. walls or dense hedges) can channel this air drainage. These can either shelter buildings from chilled airflow in winter, or lead it into buildings in summer nights. With boarded gates, you can control which it does, when.

Besides any wind-chill cooling issues, microclimate affects the air temperature around buildings. Every degree by which winter air is warmer, or summer air cooler, increases comfort outdoors and reduces heating or cooling needs indoors. Vine-covered walls trap insulating air, reducing winter heat loss by 5-20%.[7] Vines, of course, need somewhere to grow, so if you want a concrete path around your house, you can make planting wells (e.g. with plastic bowls) when you pour the concrete.

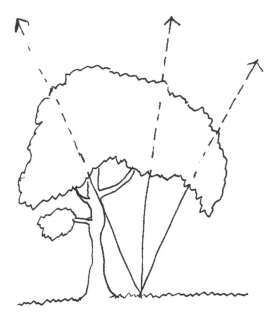

Leaf canopy reduces radiant cooling.

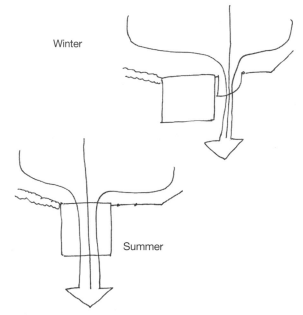

Winter

Summer

Walls/hedges and gates to channel cold air drainage:
winter, summer.

Will vines damage walls? There are three kinds: sucker, rooting-tendril and twining.[8] Wooden walls need air to keep dry so shouldn't have any sort of vine adhering to them. On masonry, different species have different effects. Sound masonry with hard mortar withstands all kinds; but sucker and rooting-tendril species can penetrate soft-mortared walls and any gaps (e.g. under windowsills, roofing-tile overlaps). Of these, English Ivy is the most destructive, so don't plant it. Boston Ivy and Virginia Creeper, however, are gentle. Whereas rooting-tendril and sucker species will climb on almost anything, twining species' growth is limited by their supporting frameworks or wires. Such frameworks should be spaced at least 50mm (2'') off walls so shoots can grip them and grow thick. This also creates airspace between vines and walls, allowing walls to dry without significantly compromising the insulation effect. For wall drying, the deeper this airspace, the better.

Will vines make walls damp? Generally, vine-covering spaced off walls keeps them *dry*. As leaves tip down, they shed water like tiles.[9] Indeed, some say that even

vines adhering to walls 'drink' (i.e. dry) more than they dampen.[10] Roofs, however, should overhang enough so that if untrimmed vines clog the gutters, overflow doesn't soak walls. Another problem with vines is that things (e.g. spiders, rats, burglars, secret lovers) can climb up them to access windows. Thorny species deter human pests, but against insect and animal pests you need insect mesh or predators (e.g. cats).

You can use climbing plants for other functions too. Wherever privacy is needed (normally at rear and side boundaries), they can cover fences. To ensure they're above eye-level, raise low fences to 1.8m (6'); the plants will probably grow 15-30cm (6-12'') higher (see: **Chapter 15: Burglar-proofing**, page 180).[11] These semi-hedges also attract songbirds. Foliage can shelter the house and garden from wind, but this means such fences must be strong enough not to blow down in gales. But will your hedge-fences obscure low-angle winter sun? If neighbouring buildings don't already do this, use annual plants (e.g. nasturtiums, runner beans, hops) or ones that shed leaves in winter along south boundaries. (Not all

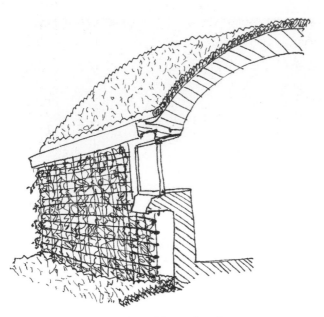

Evergreen vines on wire netting as façade.

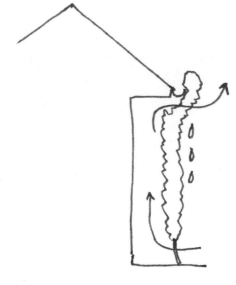

Supporting frameworks limit tendril-type vines' growth and allow an airspace so walls can dry. Roofs must overhang enough so clogged gutters don't cause wet walls.

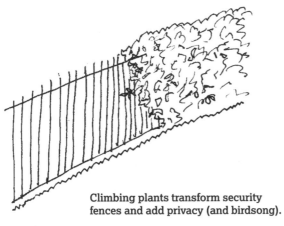

Climbing plants transform security fences and add privacy (and birdsong).

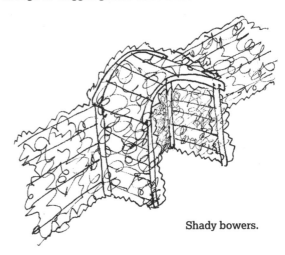

Shady bowers.

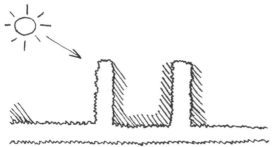

The space between bowers can form wind-protected sun-traps for spring days.

deciduous species do: privet, for instance, is ever-green, and beech hedges hold many leaves – good against wind, but bad for sunlight).

In summer, hedge shade would be welcome, but with higher sun, it doesn't cover much garden. For summer shade, you can create delightful shady bowers using short (1.2-1.8m / 4-6ft) climber-covered buttress-like trellises and wire-netting roofs – or (if vigorous roots won't be too close to house or drains) woven hazel, ash, willow or bamboo.

How you lay out your garden is no insignificant matter. Besides giving delight, it can save 10-30% energy in one go – and for trifling expense.

Keypoints

- Plant and shape trees and hedges and select species/rootstock for winter wind-shelter and summer shade.
- Use walls (or dense hedges) and gates to deflect breezes and channel cold air drainage: away from buildings in winter, into them in summer.
- Vine-cover walls to reduce winter heat loss and summer heat gain.
- Cover fences with climbing plants for privacy, security, shady bowers and songbirds.
- How you plant and lay out your garden can reduce your house's heating energy by 10-30%.

Choices

Options	For what activity/purpose?	
Reduce wind speed?	To minimise heat-loss from buildings.	
	To optimise vegetable-growing environment.	
	To maximise thermal comfort in outdoor suntraps.	
	To deposit insulating snow banks.	Where?
Accelerate wind?	To clear snow.	
	To cool/ventilate building interiors.	When (time of day, season)?
	To dry laundry, crops or damp walls.	
Drain away cool night air or impound it?	To minimise heat loss from buildings in winter.	
	To cool their interior fabric during summer nights.	
Summer sun everywhere? Or shade pools?	Warmth (and vitamin D) in cool summers.	
	Cooling in hot summers.	
Vine cladding?	**Issues:** Support; access behind for maintenance/painting; need for walls to air-dry; gutter-clog risk; access for pruning; pest access to windows.	

Resources

Microclimatic design

Barnes, L. and Bane, P. (1997) Understanding microclimate, *Permaculture Activist,* March 1997, No. 36.

Cheshire, C. (2001) *Royal Horticultural Society: Climbing Plants*, Dorling Kindersley, London.

Dawson, O. (1989) *Plants for Small Gardens*, Hamlyn, London.

Day, C. (2003) *Consensus Design*, Architectural Press, Oxford.

Dodd, J. (1989) Greenscape: 2. Climate and form, *Architects Journal*, 19 April 1989.

Dodd, J. (1989) Tempering cold winds, *Architects Journal,* 3 May 1989.

Olgyay, V. (1963) *Design with Climate*, Princeton University Press, Princeton, NJ.

Ronneberg, E. (1997) Windbreaks – tried and true, *Permaculture Activist*, March 1997, No. 36.

US Department of Energy (1988) Landscaping for energy efficient homes.

permaculturedesignmagazine.com

www.greendesignetc.net/GreenProducts_08_(pdf)/ Stoneman_Josh-Vines(paper).pdf

www.victoria.ac.nz/architecture/centres/cbpr/projects/ pdfs/literature_study.pdf

1. Dodd, J. (1989) Tempering cold winds, *Architects Journal*, 3 May 1989.
2. Cofaigh, E. O., Olley, J. A. and Lewis J. O. (1996) *The Climatic Dwelling*, James & James, London.
3. Ibid.
4. Dodd, J. (1989) Greenscape: 2. Climate and form, *Architects Journal*, 19 April 1989.
5. US Department of Energy (1988) Landscaping for energy efficient homes.
6. Reed, R. H. (1953) Design for natural ventilation in hot humid weather, Texas Engineering Experiment Station.
7. Ibid.; Dodd, Tempering cold winds; and Greenscape
8. Cheshire, C. (2001) *Royal Horticultural Society: Climbing Plants*, Dorling Kindersley, London.
9. *www.greendesignetc.net/GreenProducts_08_(pdf)/Stoneman_ Josh-Vines(paper).pdf* (accessed: 22 May 2014).
10. A building inspector's view: Oliver, A. (1988) *Dampness in Buildings*, Blackwell, Oxford.
11. If the hedge is over 2m (6'6") high and shades your neighbour's garden, however, they can complain under the Anti-social Behaviour Act, 2003.

New-build: site choice and planning

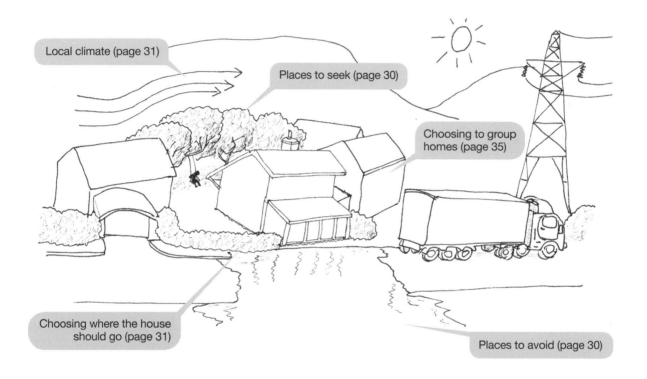

CHAPTER FOUR

Local climate (page 31)

Places to seek (page 30)

Choosing to group homes (page 35)

Choosing where the house should go (page 31)

Places to avoid (page 30)

Location basics

When choosing where to build your home, there are places to choose and to avoid. Avoid those unhealthy or risky to live in. However cheap they may be, you'll live (or die) to regret choosing them. Conversely, choosing somewhere near enough to walk or cycle – or catch a bus – to work, shops, amenities and schools will save time, bills and CO_2 many times over. You also need to consider how a place will be in other weathers and how it may change over the years.

Once you have bought your plot, where should the house go on it? You need to consider both how the house will feel indoors and how you'll use the outdoor space. This raises the issues of solar orientation, microclimate, privacy, access and space. Besides the house location, this involves any sheds or outdoor stores (if you need them) the garage and, of course, the garden. There's also the question of whether you want to, or can, share anything with neighbours and/ or friends. Sharing things, party walls and space can bring many advantages: it can save energy and money, increase amenities, green space and security, support child development and foster socialisation. Although it also brings dispute risks, there are ways of pre-empting, mitigating and resolving these.

These are all things to think about before you buy; and again before you build.

Choosing an area

There are some places never to choose to live in. Nobody should live on a flood plain (i.e. not below a 200-year flood contour, which is fast becoming a 20-year one), on toxic-polluted soil or exposed to traffic exhaust or industrial fumes. Nor should anyone live close to power lines or transformers (i.e. not within 1m (3'4")/1,000 volts of aerial cables,[1] or 60m (200') from high-voltage underground ones[2]). Nor near telecommunication masts.[3] Although the outgoing microwave beams from these are safely above head-level, incoming beams have scattered, so broadened. Frequent loud noise and vibration (e.g. from traffic, aircraft and some industry) are also unhealthy, both psychologically and physiologically. Socially, noise causes increased aggression.[4] Heavyweight walls and draught-sealed triple-glazed windows can greatly reduce outdoor noise indoors, and green roofs[†] reduce overhead noise by up to 18 decibels[†].[5] However, nothing except noise screens or distance can stop it coming in through open windows or diminish it in the garden.[6]

Among positive factors to guide your choice of site, it's convenient – and saves time, energy and CO_2 – to

be near a bus-stop and within easy, attractive and safe walking or cycling range of work, shops, amenities and schools. Good neighbours, a sense of community, a good history (for good place-spirit) and a safe neighbourhood are also important.

When you look at a site, however, you can only see what it's like *now*. It may not be the same in rush hour, on Saturday night or in other weathers. It certainly *won't* be the same in 10 or 50 years. So you should also consider how it has, does and *will* change over time. Will there be more traffic on the road? New houses blocking the view? Bigger trees blocking the sun? No trees at all when plantations are felled at maturity? The future isn't entirely unpredictable. A place's transformation journey from the past, combined with emerging pressures (e.g. demographic changes, new industries and transport links, oil depletion), gives an idea of how it will probably change.[7] If you hope to live a long time in any home you build or buy, it's important to consider this.

Keypoints

- Avoid flood plains, polluted ground, heavy traffic exposure and strong electromagnetic fields[†].
- Consider proximity to bus routes and ease/safety for cycling/walking to shops, etc.
- Use how the place *has* changed in the past to indicate how it probably *will* change in the future.

Choices

Options	Decision-making factors
Cheap or ideal area?	What are you able to change? Lifetime costs of extended travel distances.
Desirable now, or likely to improve – or deteriorate?	Price; likelihood of improvement; number and strength of factors driving improvement – or deterioration.

Choosing a site

The ideal site has all the desirable ingredients: lots of sun – especially on your house windows and in the private part of the garden – shade in summer, shelter from cold winds; and it isn't in a frost pocket. (see: **Chapter 3: Local climate basics**, page 22; **Choosing what should be where in the garden**, page 23). Additionally, it's quiet, private, has good views (which depend on your taste: nature, urban excitement – or both), and good soil (sloping gently to the south) for a garden (see: **Chapter 13: What we eat**, page 158).

Of course, there are so few perfectly ideal sites that you can't reasonably expect to have everything. Nonetheless, you can create or modify many of these ingredients, particularly sun exposure, microclimate, privacy, (selective) views and soil productivity.

Keypoints

- Seek a quiet, sunny, wind-sheltered site with good views and a reasonable level of privacy.
- Avoid severe frost pockets.
- Don't expect *everything*. Expect to need to modify the outdoor environment.

Choices

Options	Decision-making factors
Not many available sites or optimum site?	What can you improve?
Optimum for house? Or for garden?	How much do you value living outdoors? Will you grow food?

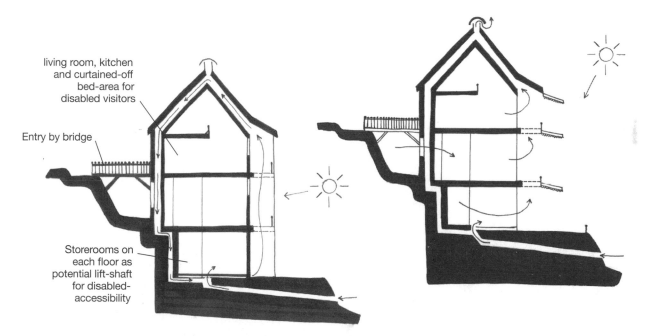

living room, kitchen and curtained-off bed-area for disabled visitors

Entry by bridge

Storerooms on each floor as potential lift-shaft for disabled-accessibility

Winter: conservatory-warmed air circulates to warm north walls and floor. Earth-warmed air inlet.

Summer: conservatory opens up to become canopy-shaded balconies. Stack-ventilation draws cool air in through earth-cooled air inlet.

Developers don't like steep land: it's more expensive to build on and doesn't lend itself to repetitive house forms. But for eco-builders, a steep south-facing slope (unshaded by the other side of the valley) is a goldmine. Moreover, as it's not desirable for development, it's likely to be cheaper.

Choosing where the house should go

Where on your plot should the house go? There are two aspects to consider: indoor and outdoor space. Sunlight through windows is a major issue. It gives warmth and raises spirits – sunless rooms are depressing. (see: **Chapter 14: Daylight and mood**, page 165; **Chapter 8: Solar heating**, page 86). Where (in which rooms, for which activities) is sunlight needed – and when (which season(s), what time(s) of day)? Along with privacy, quiet and views out, these are major determinants of where to place your house.

What about outdoor spaces? Houses in the middle of plots don't usually leave enough space on any side. Garages at the far end of gardens mean lots of hard surface (or, at least, grass-block or gravel) instead of garden – and also more snow shovelling in winter. It makes more sense therefore to keep both house and garage as far forward and to one side as possible.[8] (Proximity to boundary, however, is often limited by regulations. Also, you don't want the garage to be so prominent that it says: "only cars live here".) The wider side-space is usable as garden; the narrower one can accommodate visitor's car-parking, firewood store or sheds. This maximises usable area.

For east- or west-facing plots, houses are usually better along the north side boundary to gain as much sun as possible in the garden and on the house. To avoid wasting space, place the house as close to the boundary as is allowed by planning rules. Fire-spread regulations forbid windows too close to boundaries, so use roof-lights or light-tubes instead. North walls sheltered by overhanging roofs (if secure) make a good place to store bicycles and outdoor equipment.

For firewood, they're ideal as they're sheltered from driving rain and north winds are drier. If this firewood store, or a firewood shed, is near the vehicular entrance, you won't have to carry unseasoned firewood – heavy as it's full of water – far.

However, what if the plot faces south, so its private side – which would suit larger windows – is north-facing, so sunless? If the plot is wide enough, you can turn the house (not necessarily at a right-angle) and squeeze it to one side to allow sun through. If not, your sun-admitting windows and sunbathing part of garden must be at the front. You don't have to lose privacy, though. An above-eye-level hedge can separate a wholly private garden, deck or terrace (which might open off the living room as a summer-time extension, or off the kitchen for carrying food out) from the semi-public area (driveway, front door).

If the house is on a large plot, it's tempting to put it in the most beautiful place. Don't. Once there's a house there, it's no longer the most beautiful place. The aim should always be to make the *whole place* better: better ecologically, better microclimatically, better suited to your needs – and more beautiful.

Keypoints

- Locate the house to:
  - maximise sunlight indoors and in the garden at principal use times;
  - maximise usable garden space;
  - optimise privacy, quiet and views.
- Place garage to:
  - minimise paved/gravelled area;
  - utilise shady/less attractive ground.

Public face, private realm.

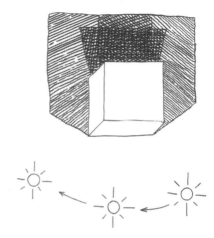
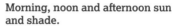

Morning, noon and afternoon sun and shade.

House in the middle of plot: useless space.

Garage at the far end of garden: lots of hard surface instead of greenery.

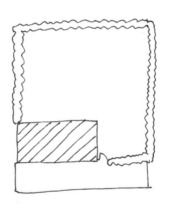

House to front and one side: maximum useful space.

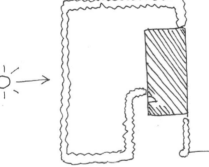

East- or west-facing plot: house along north boundary, no windows in that wall.

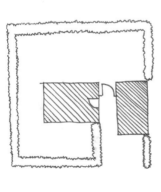

South-facing plot: house set back from the front boundary, hedge dividing semi-public area from wholly private realm.

Choices

Options	Decision-making factors
Winter sun indoors, outdoors or on solar collectors?	Use sun for spirit-uplift or heating/energy? If both, where and when is it needed?
What needs sunlight in mornings, afternoons, summer evenings? At noon? All day, all year?	How you live; what you do, when?
Garage/parking in shade, near the street or part of the house?	Need for weather-protected walk to car; fume risk indoors; effect on the street (does it seem car- or human-dedicated?); extent of paved/gravelled area; snow shovelling.

North wall firewood and cycle store.

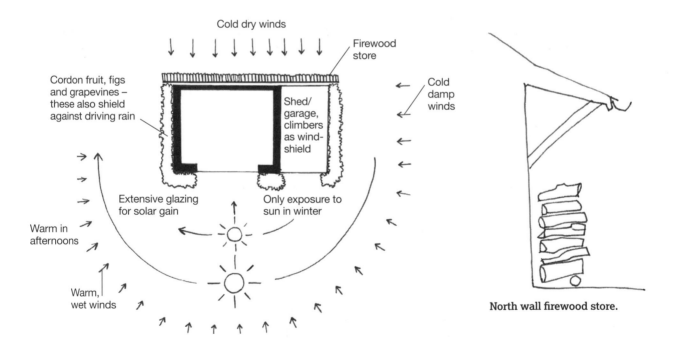

North wall firewood store.

The outsides of walls: climatic circumstance and suitable uses.

Choosing to group homes

All this about siting assumes a single, detached house on a single plot with placing limited by space-wasting regulations – as most are. But if there's an opportunity to join up houses, this is well worth considering. Joining up buildings reduces cooling surface, construction cost and the amount of land needed. The common disadvantages – maintenance and responsibility disputes, and noise transmission – can be avoided by careful contract wording and (even more careful) construction detailing.

Even better than joining up buildings, is arranging them to develop a sense of community. Shared facilities and equipment can both accelerate community formation, and reduce the environmental impacts of each home, besides saving everyone money. For example, storage, play equipment, green spaces and industrial-quality washing machines can all be shared, greatly increasing the amenities available to each household. When planning where to locate such facilities, it is important to consider how they'll be

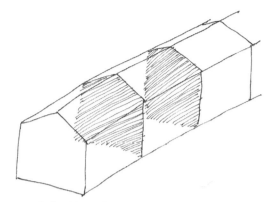

Instead of gable walls, which lose heat to the outside, terraced houses[†] have party walls to neighbours' houses at the same temperature. Besides halving wall cooling surface area, this reduces construction cost and land consumption.

used. For those that we go to frequently (e.g. laundry, recycling point, community composter) and those that focus community (e.g. meeting hall, food co-op collection-point), proximity and direct walking routes from all homes are crucial. Those that need support from the wider community to be viable (e.g. car club,

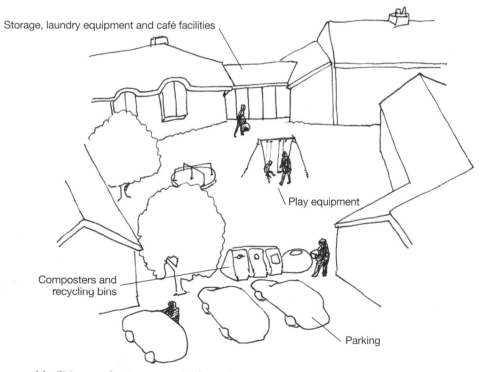

Storage, laundry equipment and café facilities

Play equipment

Composters and recycling bins

Parking

Shared grouped facilities accelerate community formation.

A communal garage can double as a noise screen.

tennis court, shop, restaurant, gym) are best sited near the entry to the estate, so they're more accessible to outsiders and don't feel forbiddingly exclusive. Technical equipment (e.g. photovoltaic and solar-thermal panels) that doesn't need attention or maintenance by individual householders is more efficient and economical if part of a coordinated design, so (like wind generators and district heating) is best communally owned. This allows greater freedom of form as every house doesn't have to have the same south-facing roof area.

Equipment-sharing typically reduces electricity consumption by 35-40%.[9] It also reduces house-building costs. A washing machine and the space in front of it, for instance, adds almost a square metre to a home's floor area, increasing construction and land cost by some £1-3,000. Sharing things, of course, raises maintenance and upkeep issues. User-recording (e.g. by card access) induces cleanliness. It's important, however, to establish a resident-led management structure to deal with such matters.

There are also benefits from sharing outdoor space. Grouping car-parking increases neighbour contact and, by de-linking parked cars from specific houses,

Communal back gardens for play ringed by homes are also ringed by adult eyes, so offer informal child supervision

makes it harder for burglars to tell who is or isn't at home. If parking is in a communal garage (albeit fenced into private bays), this can double as a noise screen from main roads. Similarly, noise-tolerant space (e.g. visitor-parking, kick-about lawns) can distance traffic noise from homes. Communal play areas encourage children's group play, which in turn fosters adult socialisation. Play areas can be 'soft', such as greens ringed by homes and/or allotment plots, or 'hard', like traffic-calmed 'home zones'[†]. Communal play areas also provide informal supervision so children can develop autonomy but still have (unobtrusive) adult eyes kept on them. These add social and child-developmental benefits to the economic and energy conservation ones.

Keypoints

- Consider joining up houses/buildings.
- Consider community-formative layout.
- Consider shared playspace for children's sociability, development and (informal) supervision.
- Consider sharing car-parking, storage, facilities and equipment.
- Ensure direct walking routes and short distances to shared facilities.
- Establish a resident-led management structure for maintenance and upkeep.

Choices

Options	Decision-making factors
What must be in every home? What can be shared?	To what extent should individual homes be fully autonomous? How voluntary or engineered should communality be? Does it feel collectivist, forced? Do lack of rules promote neighbour disputes? Or are rules unduly 'nannyish'?
Which communal facilities should be centrally located? Which should be at estate entry? Which, peripheral?	Which need quick, easy, daily access? Which require wider use/membership for viability? Which may cause nuisance?
Individual back gardens, communal back-green or home zone front?	Privacy; communality; children's play, socialisation and informal supervision; security.
Should parking be grouped? In individual house-side garages?	Do/may garages double as workshops/stores? Can communal parking ensure security? Does it encourage sociability? Double as a noise-shield?

1. Busch, H.: UK Institute of Building Biology.
2. Stakeholder Advisory Group on ELF/EMFs: UK Department of Health.
3. Although many dismiss microwave radiation as a health risk, it worries the World Health Organization and some governments (e.g. Austria, France, Germany, Sweden).
4. Research in Austria, cited in *Today* programme, BBC Radio 4, 29 May 2002; Cohen, S. and Lezak, A. (1977) Noise and inattentiveness to social cues, *Environment and Behavior*, 9, 559-572, cited in *www.monbiot.com*, 9 August 2010.
5. *www.environment-agency.gov.uk/business/sectors/91970.aspx* (accessed: 16 December 2013). Every 6 decibels reduction represents a halving of loudness.
6. Doubling distance halves noise through air. Absorbent surfaces (e.g. grass, and especially long weeds) diminish loudness. Echo, especially from hard smooth surfaces, increases it.
7. For techniques to align design with a place's transformation journey over time, see: Day, C. (2003) *Consensus Design*, Architectural Press, Oxford.
8. Garage? Mega-eco-purists don't admit to wanting one. Most other people do.
9. Roaf, S. (2010) Transforming markets for climate change – how do we do it?, *Green Building* magazine, Summer 2010, Vol. 20, No. 1.

CHAPTER FIVE

Choosing an old building to eco-upgrade

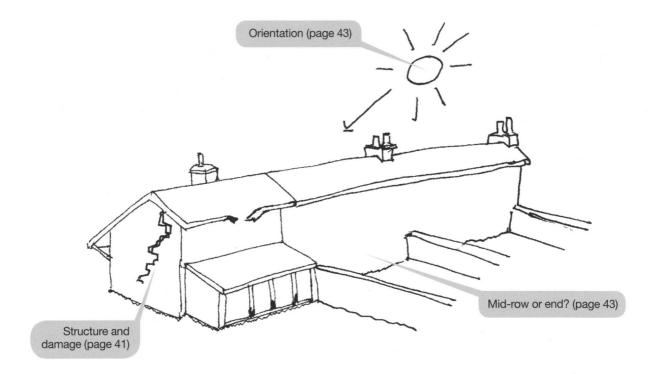

Orientation (page 43)

Mid-row or end? (page 43)

Structure and damage (page 41)

Besides the issues already covered, there are additional considerations when choosing an old building to eco-renovate. Any existing building's location, neighbouring buildings and orientation are fixed. Old buildings' construction is rarely in a perfect condition, may have a limited life and is often different from what we would choose (insulation, for instance, is rare – and there may be no room to add it). As all this limits options for eco-improvements, you need to know what you can and can't do. Before you buy a property, it's important to be able to distinguish buildings that are easy to eco-improve from those that aren't, and faults that are easy to remedy from those that mean 'don't buy'.

Old buildings basics

Old buildings may look similar to new ones, but they're different in many small but significant ways. How old is 'old'? Aesthetically, 'old' usually means pre-1850. In constructional terms, it's pre-1920. For thermal performance, however, it's virtually everything before 2006.

Wooden buildings, being vulnerable to fire, rot and insects, tended not to last as long as masonry ones. Those that still survive, however, are usually dry because wood buffers humidity. They're also often warm because, as insulation is easy to insert in the gaps between the structural framing members, this has often already been done.

In Britain, most pre-twentieth-century buildings are masonry. Unlike modern buildings, which have concrete foundations, these are built on brick or stone footings; nor did they use cement, but lime. Consequently, they're somewhat flexible. New or old, *all* buildings move with changing temperatures. Ground also moves (e.g. with moisture, prolonged loading, tree roots, vibration from traffic). If you're thinking of extending, therefore, remember to accommodate ground movement. Under old buildings, it has been loaded, so has settled, over many years. Under new buildings, it's still settling, so, when new extensions are keyed into old buildings, differential settlement can tear the two apart. To prevent this, new extensions should be tied across movement joints so they can slide down as the ground compresses under its new

load. Unless so accommodated, even half a centimetre of movement can initiate a lot of damage. Similarly, as cement is inflexible, cement-bonded repairs introduce localised rigidity, which often causes cracks – into which rain enters, but can't get out.

Another difference between cement- and lime-mortar is moisture-diffusion rate: lime absorbs water a bit faster than cement but dries out *much* faster. Most pre-1930 buildings have solid walls. Lime-mortar lets these dampen with rain, then dry out. Cement-mortar repairs, however, mean they can't dry fast enough, and only do so through the bricks, eventually destroying them. Vapour-impermeable masonry paint has a similar destructive effect. Whatever a solid wall is built of, however, driving rain can penetrate through the tiniest of cracks and leave it wet all winter. Rain-shields (e.g. slate-hanging, timber boarding) over such walls can easily remedy this. These must be rear-ventilated so that condensation dampness (and leakage) can dry out.

After 1930, cavity wall construction and cement mortars became the norm, although some post-war buildings experimented with solid concrete or concrete panels. From about 1950, breezeblock inner leaves[†] became common, and (because breezeblock is lighter than concrete block) also interior partitions. Although considered to be an insulation material, breezeblock is actually a very poor insulator. It's also radioactive (being made from foundry waste) and too crumbly to fix handrails on to (see: **Chapter 16: Physical accessibility**, page 185). Insulation blocks came next (and from 1985 were normally required by building regulations): these are more insulating but also too weak to anchor handrails, and most kinds are from similar foundry waste sources, so are also radioactive. Being in the form of radon gas, this radioactivity caused little problem when houses were draughty, but it isn't something you want in an airtight house. The breezeblock interior partitions are easy to demolish, but the exterior walls' inner leaf is part of the structure. This leaves you with a choice: seal the wall's interior surface (and accept loss of its moisture-moderating capability); build an additional solid-block inner leaf (good for thermal stability but shrinking room size); increase ventilation rate to well above

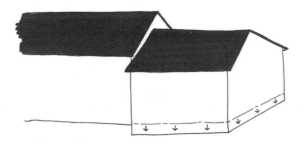

Differential settlement under an extension. Under the old part, the ground has settled over many years. Under the extension, it's still settling under its new load.

PassivHaus[†] levels (but not enough to be draughty); or buy another house.

Before the mid-twentieth century, most roofs had attics. Although sometimes obstructed by A-frame trusses, most have enough space to use for storage or as sleeping lofts. Many are big enough to use as rooms – hence the fashion for loft conversions. Trussed rafters[†], however, became common from the 1960s and 1970s. These fill the roof-space, rendering it unusable without major restructuring. Moreover, as they're made of flimsy timber sections gang-nailed together, they're prone to failure if notched or damaged. Chipboard floors also date from this era and, like trussed rafters, are still used. These give off formaldehyde, which is a health risk (see: **Chapter 13: What we breathe**, page 151).

Most more recent (British) houses have tiny rooms. If you want larger (but fewer and multi-function) rooms, it's worth measuring floor depths. Are joists[†] (floor depth less 30-38mm) sufficiently deep to span between external or party walls? If so, you can demolish partition walls to improve layout. If not, you can still demolish partitions, but will need to support floors with beams, and probably posts as well – or just put up with small rooms. Recently built houses also typically contain many more VOC[†]-emitting substances than twentieth-century houses (again, see: **What we breathe**, page 151). Being new, these are emitting at peak volume. As you've paid for all this plastic and glue, you may feel disinclined to rip it out. Or perhaps this is another type of house *not* to buy?

Most older buildings were well built, with better quality timber than that available today. In Britain, however, few were insulated, and, although minuscule insulation requirements appeared after the war and increased in the 1970s, 1985 and 1995, hardly any were insulated to current (2006 regulations) low-energy standards. Most old buildings were also draughty. Additionally, few were rodent-proof. Typically, therefore, they're solidly built but cold, often damp and sometimes pest-infested. Moreover, many have suffered depredations from time (including rot and woodworm) and bodging by ill-informed 'improvers'. Also, with age, lots of little things that we don't normally think of tend to wear out (e.g. lead flashings[†]

Keypoints

- Buildings from different times were built differently, using different materials.
- Repairs will cause damage to pre-1930 buildings if they don't take account of this.
- Lime-mortar-built solid walls dampen with rain, then dry out. Don't hinder this breathability by using cement or vapour-impermeable paint.
- Few older buildings were insulated, and most were draughty. Many have suffered depredations from time and bodging.
- Before draught-proofing, evaluate radon (and other toxic gas) risks.
- Extensions should allow for differential settlement with movement joints. They must *not* be keyed in.
- Most structural faults are due to changed circumstances.

crack, roof tiles embrittle, brass plumbing fittings de-zincify, water pipes corrode…). Like old cars, once you fix one thing, another needs attention – an unending process. Nonetheless, old houses are usually cheaper to buy, often have more character than new ones – and generally were more soundly built. If there are structural faults, therefore, it's worth asking "why?" What has changed since they were built?

Choosing an old house

Just like choosing a site for a new building, the first consideration is: do you want to live there (the area, neighbours, noise, views, local facilities, etc.); then: is it practical (near bus/cycle routes, shops, work, etc.)? Additionally, is the house attractive (inside and out), or can it be made so? Do rooms get enough daylight? If not, can this be improved?

Is it in a condition that you can live in, accept or are able to improve (and have you allowed for this in your budget)? Does its construction suit the improvements you have in mind? Is it structurally sound?

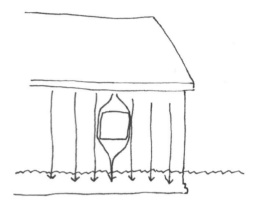 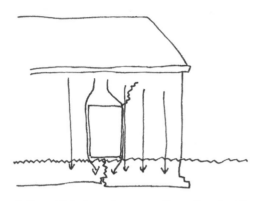

Lengthening windows into doorways can alter load distribution on foundations. This doesn't necessarily matter – but it may.

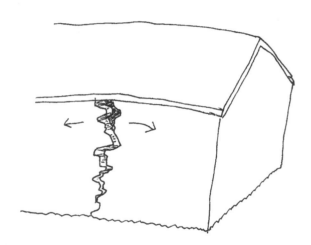

Cracks wider at the top than bottom means they're opening up.

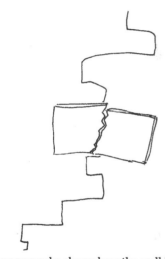

Glass glued across cracks shows how the wall is moving

Structural problems: cracks

Over the years, wood and plaster contract and build-ings settle and are subjected to various stresses, so cracks aren't uncommon, but do they matter? For this, you need to know what caused them. Common worrying causes include:

- New loads (e.g. additional floors),.
- New load distributions (e.g. converting uninhab-ited attics into rooms full of heavy furniture, length-ening windows into doorways).
- Rot- or woodworm-weakened roof-truss collars (causing roof loads to push walls apart)

- Differential settlement (e.g. under extensions).
- Soil movement (e.g. due to vibration from traffic), and soil shrinkage or swelling (e.g. due to tree roots drying clay subsoil so it shrinks; or blocking drains[1] so it's wet and swells).[2] For this reason, insurers don't like 10m (33') (sometimes even 5m/16') trees within 5m (16') of buildings.[3] Felling large trees, however, can also bring problems: as these no longer drink, soil can become wet.[4]
- Bodging: an all too common cause of structural weakening (especially of roof and floors) and of moisture ingress and condensation.

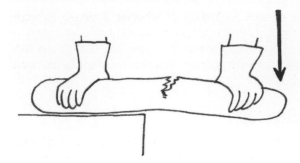

Are cracks wider at the top than the bottom? This usually means they're opening up. This is serious.

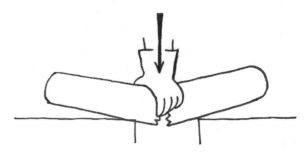

Are cracks wider at the bottom than the top? This usually means a localised support failure or localised heavy load. This is usually less serious.

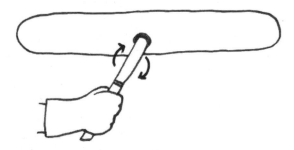

Are cracks only (or at their widest) in the middle of the wall? This is usually the result of one-time damage. This is rarely serious.

The stale baguette test: loads, supports, one-time damage and cracks.

Are such causes ongoing? Have they already caused the building to start destroying itself (e.g. do walls lean too far?)? To understand what's probably happening, you can use the stale baguette test.

Are cracks wider at the top than bottom? This usually means they're opening up. It's *possible* that you can remedy this by tying both sides together with stainless-steel wire, bolting opposing walls together with threaded steel rods and spreader-plates, or by building a buttress (or an extension that functions as one. On some soils, though, buttresses can settle, making things worse.) Perhaps you shouldn't buy this house?

Are cracks of consistent width? In gable walls, this is often the one-time result of chimney fires, so they just need filling – and perhaps stitching across with stainless-steel wire ties. Are cracks wider at the bottom than top? This suggests soil is settling, or has been washed away. *Possibly* the foundations can be underpinned or stabilised. Before even thinking of this, get professional structural advice.

If you find such cracks that your surveyor didn't notice (they may have been hidden behind dry-linings[†]; or perhaps the surveyor was drunk), you can glue a piece of glass across them. If it breaks, it'll show how the wall is moving, how fast and whether 'stitching' is feasible. If it doesn't break, the wall isn't moving any more: the crack was a one-time event.

Structural problems: rot and insect damage

The other major risk is rot. This is likely to be out of sight – but it smells. Anywhere that smells of mushrooms should ring alarm bells – especially if it's unventilated. Dry rot must be totally cut out, burnt and replaced, as it will spread to dry areas. Wet rot doesn't spread to dry areas so, while it's structurally weakening, it's not catastrophic.

Insects also damage buildings (e.g. woodworm, termites, death watch beetles: all geographically specific). You need to know whether they're still active. Is there wood dust (like fine sawdust) near their holes (e.g. on spiderwebs) or when you tap the wood? If not, they've eaten the nutritious sapwood, then left:

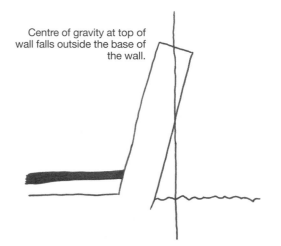

Centre of gravity at top of wall falls outside the base of the wall.

Has the building started to destroy itself?

such timber doesn't need 'treating' (meaning: poisoning). Some wood is more prone to woodworm attack than others: British-grown Sitka Spruce, for instance, makes nice poles but as it grows much faster than in the Arctic, there's more sapwood for them to feast on. They also seem to enjoy eating smaller branches of Ash. Termites, incidentally, are structural engineers: they always leave enough sound wood to hold *their* home up. If you move furniture around, however, you may redistribute the loads so the house collapses: both the termites and you become homeless.

With both wet rot and insect damage, structural integrity can be ascertained by seeing how far in a brad-awl, knife or electrical screwdriver can penetrate, i.e. how much solid wood is left. If structurally inadequate, you can splice and bolt on new joist ends – or replace in entirety. Borate-based preservatives – which are safe for mammals[5] – can be used against both rot and insects. However, for rot, the first step is to dry the area by attending to leaks and rising damp and ensuring good ventilation (see: **Chapter 9: Making an old building dry**, page 95). Also separate timber from masonry (e.g. with a damp-proof course[†] (DPC) or air gap).

None of these investigations described above can substitute for a professional survey, but they can give a general indication of whether a house is worth buying or has too severe problems for your energy or budget. Its structural state also establishes the context into which any remedial and renovation work must fit.

Sunlight

Another key issue is solar orientation. As you can't turn buildings round once they're built, this is critical. North-facing houses have south-facing gardens (good for solar-heating sun-room extensions, or just big windows). They also usually have south-facing roofs for solar panels (parapeted eighteenth- and nineteenth-century buildings excepted: their roof-ridges typically run front-to-back). Also, especially if you visit on a cloudy day (so can't see cast shade), look carefully at the garden. Will it be unduly shaded by neighbouring buildings or trees? Might these trees grow higher or are they fully mature? Will windows (and solar collectors) get enough sunlight in winter? This affects mood, enjoyment and warmth as well as energy use.

Relation to other homes

There are pros and cons to living in any type of home – from high-rise apartments to detached bungalows – and what is best for you will depend on your priorities. Bungalows are easy to make fully disabled-accessible but, however well insulated, have lots of surface to lose heat from. High-rise apartments may not sound particularly 'eco', but many have warm neighbours above, below and on both sides, so only lose heat from front and rear walls and windows. Although neighbours may cause other problems,[6] heat loss in such properties is small. With so little cooling surface, upgraded insulation can reduce their energy bills to almost zero but, as they have neither cross- nor stack-ventilation (corner and 'scissor'-type apartments excepted), can cause them to *overheat* in summer.

Terraced (joined-up) houses have only two external walls to lose heat from (besides roof, ground floor, windows and any rear extensions). Corner (and semi-detached) houses have three external walls to insulate: more expense and more heat loss. However, they also have the potential for windows in three walls. This can greatly benefit rooms, providing more alive and more evenly distributed daylight, and more views (see: **Chapter 14: Daylight and mood**, page 165). This isn't always an advantage, though: car headlights may shine straight into bedrooms, and there may be more traffic noise. Moreover, many corner buildings have smaller gardens, often with somebody else's gable wall (heightened by chimneys) at the end, which blocks sun and shortens view. Security-wise, whereas corner buildings have a street to back garden boundary to burglar-proof, mid-row houses have only boundaries to neighbours. Against this, corner buildings offer the opportunity to have doors or gates into their gardens, whereas mid-row houses have nowhere to store bicycles without bringing them (and their mud) through the house (and obstructing fire exits). Back alleys solve such problems but increase burglary risk – so you would need to raise all three garden walls with roses (see: **Chapter 15: Burglar-proofing**, page 180). The relative merits of corner or mid-row houses depends, therefore, on the value you accord to all such factors.

Resources

Energy conservation

General
Griffiths, N. (2012) *Eco-House Manual*, Haynes Publishing, Yeovil.

www.energysavingtrust.org.uk

Structure
Geo-textile root-barriers: several manufacturers

www.gharexpert.com/articles/Cracks-1468/How-To-Repair-Cracks_0.aspx

www.helifix.co.uk/applications/crack-stitching

Hutton and Rostron (specialists in environmentally-friendly timber treatment): *handr.co.uk*

www.channel4.com/4homes/build-renovate/structural-problems

www.localsurveyorsdirect.co.uk

Keypoints

- Is the house practical and do you want to live there?
- Is it in a condition that you can live in or are able to improve?
- Does its construction suit the improvements you have in mind?
- If there are structural faults, ask "why?" What has changed since it was built? Are these causes ongoing? Have they already initiated self-destruction processes? Are cracks still opening?
- Does anywhere smell of rot? If so, dry and ventilate. Cut out and burn all dry rot and replace with borate-preservative-treated timber. Check structural integrity of wood affected by woodworm or wet rot by pushing in a sharp tool.
- Solar orientation is critical: it can't be changed.
- Are windows or garden shaded by neighbouring buildings or trees? Might these trees grow?
- Well-insulated high-rise apartments lose hardly any heat but can overheat in summer.
- Mid-row terraced houses lose less heat than corner houses, but have less potential for increasing daylight. Corner houses have three external walls to insulate, but the potential for windows in three walls.
- Corner buildings often have smaller or view-blocked gardens.
- Mid-row houses' back gardens are more secure but there's nowhere to store bicycles without bringing them through the house.
- Back alleys solve such problems but increase burglary risk.

Choices

Options	Decision-making factors
Mid-terrace or end-of-terrace?	Heat loss; daylight; views; garden / bike shed access; security; headlight nuisance.
Back alley or secure rear boundary?	Security; garden / bike-shed / parking access; view length.

1. Most tree species' roots are relatively shallow, so can be intercepted with geo-textile root-barriers.
2. Big trees close to buildings and clay soils aren't a good combination. Take professional advice before felling thirsty trees, though. Clay swells when wet as well as shrinking when dry. Your building might be used to the drier soil, so stopping the tree drinking could make the situation worse.
3. Although birch trees look attractive right beside Nordic houses, their photographs don't show critical information. Foundations are below 2m (6'6") (for frost) or on rock, so roots can't damage them; moreover, cold winters and poor soil keep trees small. Shallower foundations need root-barrier protection. With shrinkable clay subsoil, keep away from trees.
4. Get professional structural advice before deciding to do this.
5. Nothing is *wholly* non-toxic. Borate toxicity is very low.
6. Indeed, you may wish 'neighbours from hell' had stayed there. A communal garage can double as a noise screen.

III. HOW ISSUES

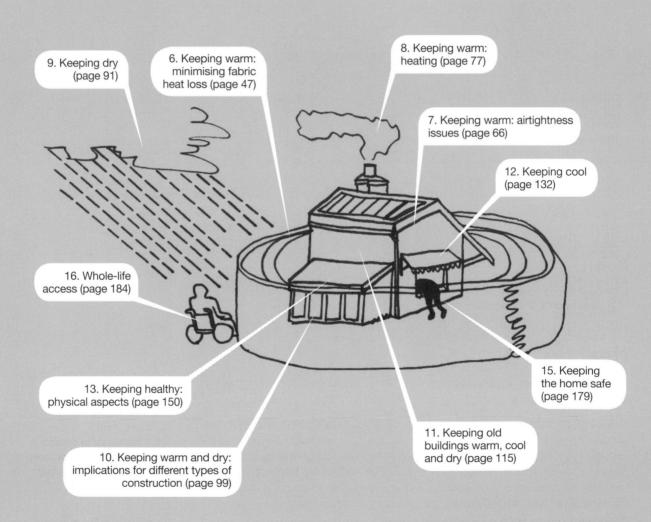

Keeping warm: minimising fabric heat loss

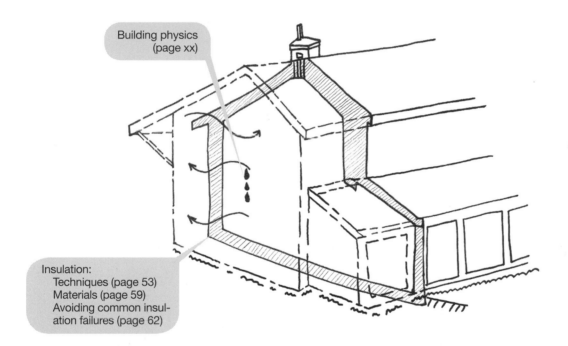

CHAPTER SIX

Building physics
(page xx)

Insulation:
    Techniques (page 53)
    Materials (page 59)
    Avoiding common insul-
    ation failures (page 62)

Heat-loss basics

There are three ways anything transfers heat to something else: convection, conduction and radiation.

'Anything' includes us. Compared to water or most solid materials, air doesn't conduct heat well, so – unlike lying on cold ground – we rarely sense heat loss through conduction to air. In cold air, however, we lose lots of heat by this means, in combination with convection. We warm the air next to our bodies by conduction, then convection replaces this air with more cold air to warm, so we're constantly drained of heat. (This is why string vests – or net curtain layers on windows or vines on walls – keep us warm: they trap air.)

Humidity can greatly affect the rate at which we lose heat through conduction, convection and evaporation. In hot weather, humid air slows sweat evaporation, so feels 'airless' and can exhaust us at some 5°C (9°F) lower temperature than in

- **Convection:** heat is transferred by particles moving within a fluid (liquid or gas, e.g. water or air).

- **Conduction:** heat is transferred within or between materials in contact with each other (e.g. between your skin and cold ground).

- **Radiation:** heat is transferred directly by electro-magnetic radiation. Materials can be distant from one another but must be in line-of-sight (e.g. you sitting in front of a fire).

dry air. Damp air conducts well: in cold moist air, we need 1-2°C (2-4°F) higher temperature to feel comfortably warm – significant as every degree increases heating bills by 10%.[1] Moist air is why British winters feel so bitter, despite relatively mild temperatures.

Dry air hardly conducts at all. Consequently, in very dry still air (e.g. in desert nights or freeze-dried Arctic conditions), there's little convective heat loss. Radiation, being unaffected by humidity, is then the principal mode of heat loss. Dry air's lower conductivity is one reason wooden and earthen houses feel warmer than they actually are: both clay and wood moderate humidity. Even thin clay plaster helps.

The other reason they feel warmer is to do with radiation. We radiate heat to cold surfaces: the colder they are and the larger the surface we're exposed to, the more heat we lose. As a result, cold building interiors (made of materials which conduct heat well, like metals and stone, rather than insulating materials like wood and straw-clay) can feel bone-chilling even if the same air temperature is tolerable elsewhere. Radiant heat loss explains why 18°C (64°F) indoor air temperatures can feel unacceptably cool in winter, but comfortable in summer. In summer, the surrounding surfaces are warmer so we radiate less heat to them.

Radiation also explains why cold rooms with insulation on the inside *feel* warmer than those with thicker insulation further out in the construction. Paradoxically, even though they actually *lose more heat*, this is so even if the insulation is thin, like timber panelling, hanging rugs or cork pin-boards.

Indeed, thermal comfort and minimum building heat loss aren't always the same. Our feet, for instance, feel warmer on a thick carpet over an uninsulated floor than on a steel floor on 300mm (12") of insulation. The steel conducts body heat to a large surface, which is then cooled by room air – cooler than us – making *our feet* feel cold, although the actual heat loss *from the building* is negligible.

In cold environments, our main concern is keeping warm: keeping *ourselves* warm. The actual building temperature is less important – and much more expensive to optimise. By far the cheapest and least environmentally harmful way to minimise heat loss isn't to insulate buildings but to insulate *ourselves* with warm clothing – as did our ancestors. (Being designed to keep animals warm, wool, fur, down and silk make the warmest clothing materials.) Keeping warm starts with clothing, but extends to clothing the things we touch (e.g. carpeting floors and fabric-covering chairs) to produce low-conductivity surfaces. Traditionally, sofas had skirts and tall backs to protect ankles from draughts and to stop heat radiating from our backs to cold surfaces behind us.[2]

The same principles apply for buildings as for heating bodies, but the practice is different. Buildings lose some heat by conduction to ground, a little by radiation (especially through windows) but most by convection (after conduction to air) – especially from the roof (more exposure to wind and, as hot air rises to ceilings, greater temperature difference). As the ground temperature is usually closer to indoor temperatures than is winter or summer outdoor air, 'earth-coupled' rooms (e.g. basements, earth-sheltered buildings, cave-dwellings) tend to maintain stable temperatures. 'Sky-coupled' rooms (e.g. penthouses, attics, glasshouses, buildings on stilts), on the other hand, respond rapidly to outdoor temperature changes.[3]

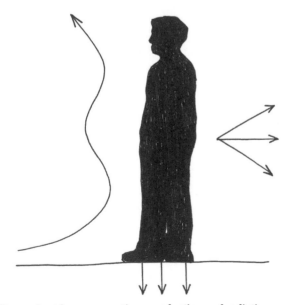

Human heat loss: convection, conduction and radiation.

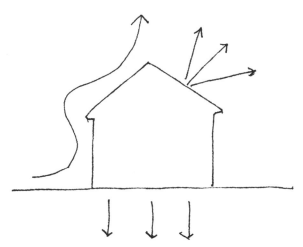

Building heat loss: convection, conduction and radiation.

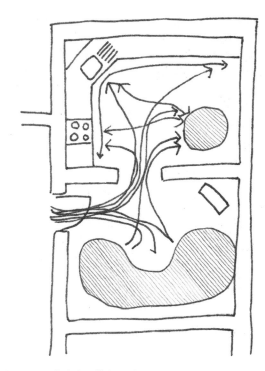

Routes and sitting/doing places.

Heat loss is proportional to cooling surface area. (This is why the typical Inuit body shape is small and rounded: less surface area and a shell of insulating fat.) Smaller houses have less surface area in total (although their surface-to-volume ratio is higher) and also less air volume to heat; they therefore have lower heat loss. (Hence poorly insulated mobile homes can be cheaper to heat than super-insulated mansions.) Additionally, smallness typically reduces all ecological impacts.

To reduce size without compromising usability, it's important to distinguish places for sitting and/or doing things in from routes between them. The more destinations each route serves, the less floor area is needed for 'circulation space'. Additionally, thinking in three dimensions often reveals wasted spaces that can be used. Bed lofts, for instance, free a lot of floor space, sometimes even giving minute rooms more free floor area than much larger rooms. Especially for small rooms, non-rectangularity (in plan and/or ceiling angle), large low-silled windows and visually softer surfaces (e.g. wood, hand-finished plaster) make them less claustrophobic. More about this later (see: **Chapter 14: Spatial factors**, page 167).

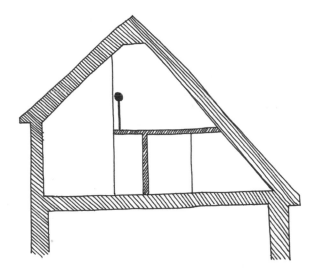

Bed lofts free floor space.

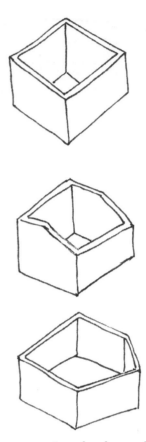

For small spaces, non-rectangular plans and/or sections are less claustrophobic.

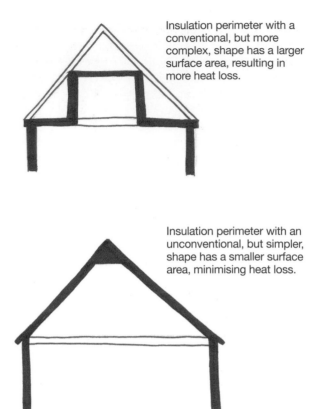

Insulation perimeter with a conventional, but more complex, shape has a larger surface area, resulting in more heat loss.

Insulation perimeter with an unconventional, but simpler, shape has a smaller surface area, minimising heat loss.

If the insulation perimeter is within the building, try to minimise its extent: this reduces heat loss, (potentially) increases usable space and (usually) results in less vapour-check sealing fiddles.

Simpler forms have lower surface-to-volume ratios than complicated forms. If a building's form (or supporting it on stilts) doubles its 'skin' area, it needs twice the insulation of a compact building. The most compact form is a sphere – which is good for a foetus to live in but is somewhat impractical for houses, as spheres lack flat floors. Hemispheres (e.g. igloos), cones (e.g. tipis, kåtan) and cylinders (e.g. yurts) not only minimise cooling surface and shed wind, but also reflect body- and fire-heat back onto occupants. These, however, don't suit manufactured rectangular furniture so, although domes were a 1960s fashion, rounded forms don't appeal to volume house builders.

Although hemispherical (or cubic) buildings lose least heat, long east–west axis buildings get the most winter sun – so gain the most solar heat. Buildings lose more heat than they can gain from the sun, meaning that energy conservation favours compact forms. However, as buildings are built to be lived in, we also need sunlight to lift mood and for physical health (see: **Chapter 14: Daylight and mood**, page 165). The optimum house form, therefore, must balance compactness to reduce heat loss against maximised south façade area for solar gain – and of course, other issues, like *over*heating risk; and views, daylight, room shapes, etc…. (You could build a cylinder, one-third glazed, two-thirds heavily insulated, which rotates with the sun. Given the expense, motor energy, complications and maintenance, though, the value of such a design is arguable – especially as there are easier methods of solar heating.) In short, there's no one optimum: it depends on site, climate, circumstance and your preferences. In designing your eco-home, insulation (and surface minimisation) to reduce heat loss, and using sunlight (and maximised south glazing) for heat gain must always be major considerations, although never the *only* determinants of your design (see: **Insulation**, page 53; **Chapter 8: Solar heating**, page 86).

All heat loss is proportional to temperature difference. Consequently, besides maximising insulation and minimising surface area, reducing indoor temperature reduces heat loss. There are three acceptably comfortable ways to do this: heating more by radiation than convection, thermal zoning and – of course, and cheapest – wearing warmer clothes.

Lowering indoor temperatures reduces the temperature differential to outdoors, so less heat is lost. Additionally, the lower the indoor temperature, the greater the contribution that incidental heat gains (e.g. occupants, cooking, electric equipment, sunlight) make to home heating. Up till 1970, British indoor winter temperatures averaged 13°C (55°F),[4] but clothing, carpets, low-conductivity surfaces and limiting heating to a single room kept occupants (just) tolerably warm. Nowadays, we expect more effortless comfort and greater personal independence, so we heat our buildings: every room in them.

The higher the indoor temperature, however, the greater the indoor–outdoor temperature differential, so the more heat you'll need to add. Part of this heat

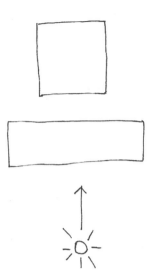

Cubic or long: less heat loss or more solar heat?

is 'free':[5] incidental heat gains (e.g. from people, appliances and sunlight). These incidental heat gains reduce the amount of heat you need to buy to that proportional to the difference between the semi-warmed indoor temperature and the chosen indoor temperature. This is a much smaller figure, so small increases or decreases in chosen indoor temperature produce disproportionately large effects on heating bills.

Heat loss isn't uniform. Think about a bowl of soup. Like soup cooling faster in thinner bowls than thick ones, windows lose heat faster than walls, so windows need additional insulation for winter nights (see: **Chapter 10: Warm windows**, page 105). Just as hot soup is cooler around the edges, underfloor insulation loses most heat from around its edges. Soup also

Indoor temperature, incidental heat gains and heating[6]

Incidental heat gains added to outside air temperature	Heating required (proportional to the temperature required)	Indoor temperature
(Say) 12°C (54°F) averaged over the heating season	7°C (12°F)	19°C (66°F)
	10°C (18°F): Almost 50% more	22°C (72°F)
	12°C (21°F): Almost double that for 19°C	24°C (75°F)

Typical temperature differences

(Figures are geographically and temporally variable, very approximate and rounded for simplicity, but illustrate insulation thickness needed)

Where		Temperature	Temperature difference		Relative thickness of insulation needed[7]
Indoor air @ +20°C (68°F)	Outside air	-10°C (14°F)	30°C	54°F	6
	Outside air with wind-chill	(effectively) -20°C (-4°F)	40°C	72°F	8
	3m (10') underground under edge of building	Approximately annual average[8] = +10°C (50°F)	10°C	18°F	2
	0.5m (1'8") underground under centre of building	Between annual average and indoor temperature = +15°C (59°F)	5°C	11°F	1

cools faster outdoors in a breeze (or if you blow on it) than in still indoor air. Similarly, house surfaces exposed to wind cool faster than sheltered ones. Indoors, warm air rises, so upper parts of buildings have more heat to lose. Outdoors, at high levels, there are fewer obstructions to slow wind, so these same upper parts are also exposed to more cooling. Sunless sides of buildings and those exposed to cold winds lose more heat than sunny, sheltered sides. Consequently, although the thickness of insulation is usually uniform, the thickness actually *needed* varies with location.

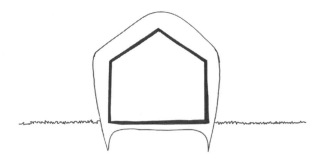

Relative amounts of insulation needed – although for easier installation, don't taper insulation. (As you've bought it, you might as well use, not waste, it.)

Not all rooms, or parts of rooms, need to be equally warm. For example, a boot lobby, or storeroom, can be cooler than a living room; the circulation area of a room can be cooler than a focal hearth area. Conventionally, rooms we're active in are about 2°C (4°F) cooler than those we mostly sit in.

Design can improve all of these heat-loss-rate factors: better and/or thicker insulation, less surface area, less heated volume and lower indoor temperature, particularly adjacent to north walls.

Basic heat conservation

- Reduce heated-space and cooling surface.
- Maximise insulation.
- Movable insulation (e.g. shutters, curtains) to windows.
- Draught-lobbied and wind-sheltered entrances.
- Eliminate uncontrolled ventilation.
- Zone warmest areas in the centre of the building or by warmest walls; coolest areas by coldest walls.
- Reduce air temperature by increasing the radiant component of heating.
- Further reduce temperature by wearing warmer clothes – this is the most effective and economical measure of all.

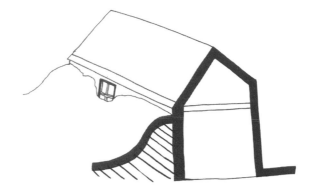

Earth sheltering (on north walls) can quarter temperature differential.

Choices

Options	Decision-making factors
Greater exposure to sun or more compact form?	Solar-heating potential; winter shading; soul-need for sunlight.
Less movement space and more 'place' space indoors or a larger home and larger heating bills?	Design.
Smaller surface area or thicker/better insulation?	Design.
Lower indoor temperatures or less cooling outdoors (e.g. warmer air, less wind-chill, less radiant cooling)?	Microclimatic landscaping; clothing; radiant walls and/or floors; thermal zoning; comfort.

Insulation

Insulation doesn't *prevent* heat loss; it *slows* it. Heat always flows from warmer to cooler, but the lower a material's thermal transmittance, the slower this journey, so the less heat is lost in any given time. So, if (dry) soil has one-fiftieth of the insulation value of mineral wool but is fifty times as thick, it'll slow heat transfer exactly as much. In the short term, walls that rely on soil insulation (e.g. basement walls) feel colder because the insulating soil is still warming up, but after a few months, there'll be no difference – except greater thermal stability. This means that thickness can compensate for poor insulation quality – and good insulation quality for thinness. Consequently, wherever space doesn't allow adequate (or any) insulation (e.g. in renovation), other (poorly insulating) material can compensate if it's thick enough. Earth sheltering (particularly on north walls, where there's no need for big windows for sunlight), for instance, halves indoor–outdoor temperature differential – or, in northern USA and Eastern Europe, quarters it. It also eliminates wind-chill.

So how much insulation do buildings need? The short answer is as much as reasonably easy and affordable.[9] There is, of course, a limit. There are constructional, structural and cost implications; and for CO_2-reduction, there's a point when the CO_2 produced by manufacturing the insulation outweighs that saved by reduced heating need. Also, proportionally, as heat lost from the building fabric declines, that lost through ventilation rises, so increasing insulation delivers diminishing returns. After a certain point, it's more productive, therefore, to address ventilation heat-loss (see: **Chapter 7: Ventilation control**, page 74). As restricted ventilation brings complications, however, this point only comes after you have insulated *very* well.

For insulation thickness, a crude rule of thumb is 1, 2, 3:[10] one unit of insulation value under floors (low temperature differential); two for walls, and three for roofs

A material's thermal resistance is called its 'R-value'[†]. This is measured in $m^2.°K/watt$. (°K, here, meaning 'temperature difference in °C'; and '.' meaning 'multiplied by'.) More commonly referred to, however, is its thermal transmittance, called its 'K-value'[†]. This is measured in watts/m.°K. The lower the K-value (or the higher the R-value) is, the better it insulates.

Most building elements (e.g. walls, roofs, floors) are constructed of several layers of materials and vary in thickness. Their heat-loss rate per square metre is called their 'U-value'[†], and is measured in $watts/m^2.°K/hour$.

An element's heat-loss rate, therefore, is proportional to: temperature difference x U-value. And for its heat loss over a year, multiply this (after averaging temperature difference) by 24 x number of heating-season days.

Whereas K- and R-values are useful to choose materials, U-values (x surface area x temperature difference x time) indicate how much heat a whole element (e.g. window, wall or roof) will lose.

For compliance with building regulations, manufacturers will usually calculate U-values for you, but if you're technically inclined, you can do it yourself:

U-value = 1_____

   1/k1 x d1 + 1/k2 x d2 + 1/k3 x d3, etc. + surface resistances and reflectances

(d1, d2, etc. = depth of the constituent layers; k1, k2, etc. = K-values of the constituent materials)

or =   1_____

R1 x d1 + R2 x d2 + R3 x d3 etc. + surface resistances and reflectances

Just to make it even more complicated, K-values are also called 'lambda values' (λ); and mustn't be confused with °K. (degrees Kelvin[†]): a measurement of temperature whose scale starts at absolute zero (0°K = -273.15°C). An increase in 1°K is the same as an increase in 1°C – the degrees have the same magnitude as each other but are zeroed at different points. For our purposes they're interchangeable, since we're talking about temperature differences. Also, American values are in imperial units (e.g. Btu, °F, ft^2).

The really important thing to know is that the larger the R-value or the smaller the K-value and U-value, the better the insulation.

Common U-values: typical UK construction

(Shaded methods meet BRESCU 'zero heating' standard[12])

Element	Construction	Typical U-value
Walls	Uninsulated brick/block cavity wall (typical pre-1970s in UK)	1.5-1.6
	Brick, 50mm (2") cavity, insulation block (typical of 1970s in UK)	1.0
	Brick, 50mm (2") cavity, insulation block; full-fill insulation in cavity	0.4
	Concrete block, 170mm (7") mineral wool insulation in cavity, concrete block	0.2
	Concrete block (or brick), 250mm (10") insulation in cavity, concrete block	0.14
Roofs	Uninsulated pitched roof	1.9
	Roof with 500mm (20") cellulose fibre insulation	0.08
Ground floor	Uninsulated solid floor	1.0
	Solid floor with 50mm (2") polyisocyanate	0.25
	Solid floor with 300mm (12") expanded polystyrene	0.10

(greater wind-chill exposure and radiation to night sky) – the 'unit' being as much as is practicable. Calculations are, of course, better than rules of thumb, but other limitations often render these unnecessary. For example, insulation thickness is usually limited by other elements, such as cavity ties for walls.

Before expending money and effort greatly increasing insulation thickness, upgrade weak points: particularly windows and cold bridges[†] (see: **Chapter 10: Warm windows**, page 105; **Insulation materials**, page 59; **Avoiding common insulation failures,** page 62). It's also worth maximising free insulation: snow, earth, air-entrapping vines and windbreaks (see: **Chapter 3: Choosing what should be where in the garden**, page 23).

Snow on roofs, or trapped against walls by shrubs or climbing vines, is worthwhile insulation. Although its insulation value varies greatly according to crystal type and compaction, it can be two-thirds as good as baled straw. Snow, of course, isn't always there; and it can cause problems. If it accumulates, it can be destructively heavy. Worse, where roofs overhang walls so are cooled above and below, semi-thawed snow can freeze into ice dams, which trap slush ponds. If this water penetrates a tiling lap, it causes disastrous leaks. Design must consider such structural and constructional implications and also 'avalanche' risks.

Despite such complexities, in most common circumstances and within reason, the more insulation on roofs, floors, walls and windows, the better. Insulation, however, affects warmth, which affects the water content of air. Consequently, to avoid problems with condensation, before installing insulation you need first to understand a little about building physics.

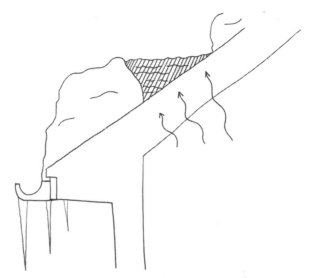

Re-frozen semi-thawed snow can trap water ponds. Impermeable sheet (e.g. zinc) is needed in risk area.

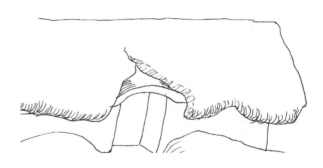

Snow insulation: avalanche risk. Snow guard or deflecting roof shape over door.

Snow: free insulation.

Choices

Options	Decision-making factors
How much insulation?	Constructional feasibility; value for money/CO_2.
Thicker/better insulation or longer heat-loss routes?	Constructional design.
Snow insulation?	Load on structure; risk of leaks from ice-dammed melt-ponds; avalanche risks.

Building physics

The higher the air temperature, the more water vapour it can hold. (Hence over kettle spouts, steam often isn't visible within the first centimetre. See: **Chapter 9:ampness and dryness basics**, page 91.) Conversely, the more water vapour in the air, the lower the temperature at which it condenses. Indoors and out, the temperature at which condensation occurs is called the 'dewpoint'. The amount of vapour in the air (called 'vapour pressure'[†]) determines the vapour's movement. Like heat, it always flows from high to low. Indoor air is often warmer than outdoor air, so it can hold more water vapour – and, from occupants' breathing if nothing else, there's usually lots of water in indoor air. This means there's always water vapour (from indoor air) in vapour-permeable construction (except in dynamic-insulated buildings; see: **Chapter 7: Dynamic insulation**, page 68).

If it's colder outdoors than indoors, temperature declines across the insulation thickness. Consequently, the dewpoint temperature can only be somewhere *within* the building fabric.

This is rarely a problem in masonry construction, as brickwork absorbs this water and then diffuses it. In wood-frame construction, however, condensation here can do a lot of damage if it can't dry out. Conventional modern practice is to attempt to stop moisture-rich indoor air from getting into the insulation, leaving only outdoor air – which has already done

its condensing outdoors when it was cold – within the insulation. In wood-frame roofs and walls, therefore, it's conventional to use a vapour-barrier. Safely on the room side of the dewpoint (so near the interior surface, i.e. within the first 10% of the insulation), this stops indoor air entering the construction – in theory! In practice, however, membranes are easily punctured – and often badly sealed. Impermeable insulation also theoretically stops air movement – but, similarly, it's easily compromised by gaps where panels join. (Moreover, foamed plastics are polluting to manufacture – and some pollute indoor air. Nor do they let walls breathe. See: **Chapter 13: What we breathe**, page 151; **Chapter 7: Ventilation control**, page 74**.**) Another problem is that hot, humid summer days and cool building interiors reverse this theoretical condensation model.

Trying to stop moist air from getting into the construction is tricky and can cause problems if it fails. It's safer, therefore, to ensure moisture can get *out.* For this, there's the 5:1 rule: materials' vapour permeability must increase towards the wall exterior, the outermost layer being at least five times more vapour-permeable than the innermost. In timber-framed walls, this 'outermost layer' is best protected by a rear-ventilated rain-screen. In roofs, the

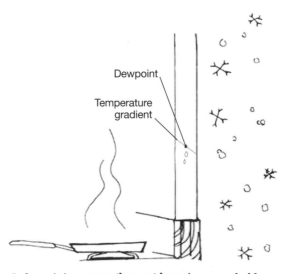

Indoor air is warmer than outdoor air, so can hold more water vapour. As temperature declines across the insulation thickness, condensation often occurs somewhere within the building fabric.

Impermeable internal surface: surface condensation.

Impermeable external surface: condensation within the building fabric.

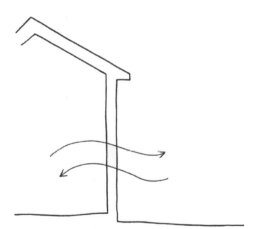

Vapour-permeable construction allows walls to breathe.

'outermost layer' is best below the ventilated air-space, protected by the roof. The innermost layer shouldn't be 100% impermeable so you should use, for example, bitumen-impregnated paper, plywood, OSB3† or hardboard† – not polythene. Even better, use an 'intelligent' vapour-control membrane; these become more permeable with warmth, so walls and roofs can dry from their indoor faces in summer.

The thicker the insulation, the deeper in will any condensation occur – so the less easily can the construction dry out. Wherever this is a risk, it's important to have a way for the water to escape. This could be a drip-drainage zone at the bottom (e.g. in cavity walls), or a drainage layer (e.g. expanded clay granules below straw-bale walls). In wood-frame walls, it's much the safest to use natural insulation materials (such as sheepswool, hemp, straw, cellulose fibre or wood-fibre). These absorb occasional condensed water then diffuse it. Otherwise, it can drain into puddles, which can stay there until the wood rots.

Building physics is a precise science, which it's irresponsible to ignore. Real-world conditions, however, can differ so much from laboratory conditions that it's wise to design as much tolerance to condensation into your constructional system as you can. This means maximising its potential to *dry out*.

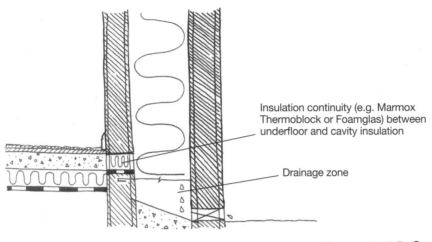

Drainage zone at the bottom of cavity.

Insulation continuity (e.g. Marmox Thermoblock or Foamglas) between underfloor and cavity insulation

Drainage zone

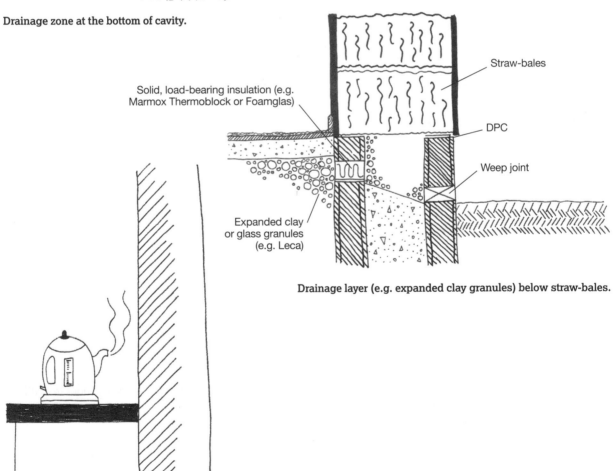

Solid, load-bearing insulation (e.g. Marmox Thermoblock or Foamglas)

Straw-bales

DPC

Weep joint

Expanded clay or glass granules (e.g. Leca)

Drainage layer (e.g. expanded clay granules) below straw-bales.

Hygroscopic materials absorb, then diffuse water vapour and so moderate humidity.

Insulation materials

There is a confusingly wide range of insulation materials. Which is best? The question, actually, is: best *for what*? Some are cheaper; some insulate better; some are much more environmentally damaging than others; some emit toxic gasses, especially in fires. This is complicated by manufacturers' unwillingness to describe their products' shortcomings. As a general guide:

- Mineral fibres are cheapest (but as glass fibres are sharp and can break into inhaleable lengths, I avoid glass fibre).
- Foamed plastics are better insulators but more environmentally damaging (although polystyrene, like polythene, is available as a recycled product) and may emit toxic gasses.
- Natural materials are safer both for home occupants and to the environment – but are often more expensive.
- The most efficient of all (e.g. vacuum panels, aerogels) are also by far the most expensive.

A further question is which is the best form: quilt, board or loose fill? This, of course, depends on where and how the insulation will be installed.

- Quilts are easy to unroll and can be cut to length in situ. They should be packed tight into gaps to prevent draughts. If you're using multiple layers, stagger their ends and edges.
- Boards are easy to fix but the structural elements you fix them to may move over time (e.g. timber shrinks with drying out and masonry moves with temperature changes), so need draught-sealing with tape to prevent thermal bypass[†].
- Loose fill is usually blown in by a specialist contractor. It fills all accessible voids so eliminates draughts. It can be applied to vertical surfaces but may settle over time.[13] This can cause cold bridges at the tops of walls unless there's an overlapping insulation layer (e.g. insulated ceiling-shaping or coving).

Types of insulation (general principles)

Type	Advantages	Disadvantages
Organic (natural origin) materials	Non-toxic. Hydroscopic, so absorb and diffuse water vapour. Good insulation value.	Rot if soaked for long.
Mineral fibres	Don't decompose (but may be damaged) if wet. Good insulation value.	Lung/throat/skin irritant. Binder may off-gas. Non-hydroscopic, so condensate forms pools of water.
Foam plastics	Don't decompose (but may be damaged) if wet. Highest insulation value.	Toxicity concerns. High embodied energy/pollution. Waste is non-biodegradable.
Glass/clay beads	Don't decompose nor suffer damage if wet. Inert.	Lower insulation value. High embodied energy.

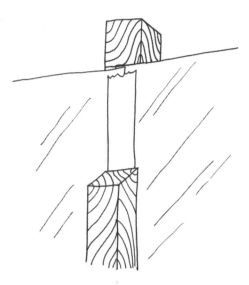

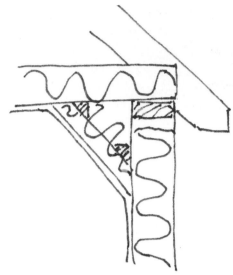

Wind-barrier and vapour-barrier or board joints should be tape-sealed and trapped under battens.

Loose fill may settle over time. Insulated ceiling-shaping or coving can prevent cold bridges at the tops of walls.

Common insulation materials[14]

Material	Source	Form	Embodied energy GJ/m³	Embodied pollution	Toxity risk to occupants	Thermal con-ductivity W/m.ºK
Sheepswool	Ex-living materials	Roll	0.11	Negligible	Negligible	0.038-0.039
Wood-fibre		Board or batt				0.038-0.043
Hemp-fibre		Batt				0.039
Hemp-lime (hempcrete)		Infill				0.05-0.07
Cellulose fibre		Loose fill or batt	0.48			0.038
Cork		Board or loose				0.04
Straw (bales)		Construction material				Approx 0.2
Straw-clay		Infill				Approx 0.6

Glass wool	Mined materials	Roll, batt or loose		Moderately high	Sharp fibres	0.044
Mineral wools (other)		Roll, batt or loose	0.83		Irritant fibres	0.042
Foamed glass		Board	2.7	High	Negligible	0.043-0.048
Expanded clay (Leca)		Loose		Moderately high		0.1
Cellular clay bricks[15]		10-49cm wide bricks				0.07
Vermiculite		Loose			Possible asbestos contamination	0.062
Phenolic foam	Petroleum-based plastics	Board or spray	Average 4.05	Very high	Concern; high concern in fires	0.021-0.038
Polyurethane foam		Board or spray				0.02-0.026
Polyisocyanate		Board			Concern;[16] high concern in fires	0.02
Polystyrene: expanded and extruded		Board or beads				0.03
Vacuum panels	Manufactured products	Board		No information	No information, but presumably low	0.007
Multifoils[†]		Roll				Up to R 4.8 m².°K/W for 100mm thickness; but manufacturers' claims are disputed. Multifoils are better insulators against solar radiation (high temperature differential) than cold weather (moderate temperature differential).

To return to the original questions: which insulation material is best? And in which form? This, of course, depends on circumstances. You probably want the healthiest to live with and least damaging to the environment, but also the best insulation for the least expense: not necessarily the same! For the whole environment, natural is naturally best – and healthiest.

If space is limited, however, you may need a more efficiently insulating material (which costs more and has greater environmental impact) so it can be thinner. Similarly, for underground applications, you may need the most durable. Which type you choose will, therefore, reflect your eco-aims, but often have to be modified by situation and budget.

Keypoints

- Vacuum panels and aerogels are the best insulators but very expensive, so only justified where space is severely limited (e.g. insulating existing ground floors).
- Plastic foams are the next best insulators but the most environmentally damaging in manufacture, may emit toxic gasses indoors, and most kinds are extremely hazardous in fires.
- Mineral fibres are generally the most economical insulation materials. (Glass fibres, however, are sharp: a possible health hazard.)
- All clay products are inert.
- Expanded clay and glass recyclate are inert and suitable for underfloor use.
- Natural, ex-living materials are healthiest to live with and have little embodied energy and pollution. They range in price from economical cellulose fibre to sheepswool (preferred by sheep for countless millennia).

Choices

Options	Decision-making factors
Mineral fibre, plastic foam or natural insulation materials?	Health risks from breathable fibres, sharp dust, toxic off-gassing, smoke if burnt; environmental impact; efficiency/thickness; durability in adverse conditions; effect on indoor air quality; vapour-breathability; condensation risk.
Quilt, board or loose fill?	Ease of application; draught / thermal bypass[†] risk due to poor fit.

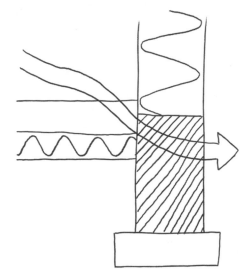

Avoiding cold bridges: draw heat-loss routes on to construction drawings. Then revise the construction.

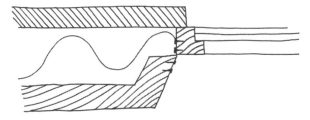

Avoiding cold bridges: all insulation, door- /window-frames in one unbroken layer.
Avoid bridging the insulation layer.

Avoiding common insulation failures

Good insulation means warm interior surfaces, but cold bridges provide cooler points, lines and/or areas on which condensation, then mould, can appear. The heat loss isn't usually serious, but the mould is. Although it's probably impossible to insulate potential cold bridge points to the same standard as the other construction, it's therefore vital to at least diminish the leakage of heat. Avoiding (or minimising) cold bridges needs very careful design. The easiest way to identify places where insulation is lacking, too thin, bridged or bypassed is to draw the heat loss routes on to construction drawings. The surest system of preventing cold bridges, however, is to keep all insulation, door- and window-frames in one unbroken (or overlapping) layer. In neither plan or section should there be any breaks. Try also to minimise steel fixings across the insulation layer. For super-insulation, even this isn't enough. PassivHäuser also overlap window- and door-frames with 60-70mm (2½-2¾") insulation.[17]

The better the insulation, the more critical cold bridge avoidance becomes. In super-insulated buildings, unexpected things become cold bridges. For example, brick walls with wide insulation-filled cavities require thicker cavity ties for strength, but these are more heat conductive than the thinner ties used in less-insulated walls. You can overcome this by using epoxy-bonded basalt-fibre cavity ties (0.7W/mK thermal conductivity) or, less efficiently, polypropylene or high-tensile stainless-steel wire. These allow wall cavities to be up to 300mm (12") wide for insulation. Similarly, timber rafters and wall studs[†] conduct significantly more heat than the insulation between them. This can be overcome by using engineered-timber I-beams (or double-stud wall construction) instead of solid timber, and/or by lining the interior faces of walls and ceilings with insulation board.

One unanticipated heat-loss route, which all too easily can occur, is 'thermal bypass': draughts within the construction that (although not necessarily entering rooms) bypass the insulation. This can reduce insulation effectiveness by 40-70%.[18] To prevent thermal bypass, construction must be wind-tight. In timber construction (e.g. roofs, wood-frame walls), wind-proof, but breathable, membranes outside the insulation prevent this.

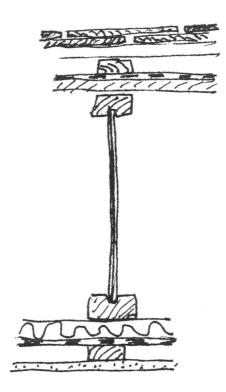

Engineered-timber I-beam studs and rafters, lined with insulation board against cold bridges. Cable void ensures services[†] don't puncture vapour-check membrane.

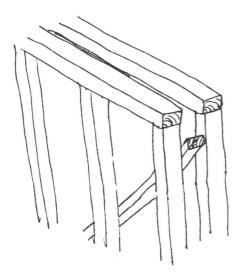

Double-stud wall construction.

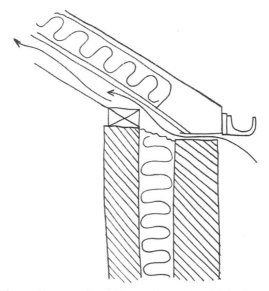

Thermal bypass: draughts can draw cold air into the building.

Just as for cold bridge avoidance, the easiest way to prevent thermal bypass is to draw likely airflow routes onto the construction-section drawings. For cold bridges, you're looking for ways to slow heat transfer (with insulation) or lengthen its route through poor insulants (e.g. soil, masonry); for thermal by-pass you need to trap membrane overlaps so they don't open when sealing tape loses its adhesion (see: **Airtight construction**, page 70). Such little details make a big difference to insulation efficacy.

In designing to prevent heat loss you need to:

- avoid cold bridges;
- seal draughts (besides the above, see: **Chapter 7: Airtight construction,** page 70);
- ensure ventilation of the spaces above or outside the insulation (see: **Building physics,** page 56).

All three are important. However, if space, access or sequence prevent you from achieving them all, which is the *least* important?

Failure of	Physical consequence	Result
Ventilation above/outside the insulation	Condensation within the fabric, potentially causing mould spores and rot, both inaccessible for treatment.	Disaster
Cold bridge avoidance	Localised condensation and mould.	Seriously unhealthy
Draught-sealing	Heat loss and localised draughts.	Expensive and uncomfortable

Personally, I prefer discomfort to disaster.

Keypoints

- Prevent cold bridges by keeping all insulation, door- and window-frames in one unbroken layer.
- Prevent thermal bypass by wind-tight construction or a wind-proof membrane outside insulation.
- Use low-conduction materials and/or construction systems.

Choices

Options	Decision-making factors
Single insulation layer or overlapping insulation?	Building form; location of structural elements (e.g. posts); depth of window recesses.
What fixings through insulation?	Availability of low-trans-mittance fixings.
Wind-tight construction or wind-proof membrane?	Compressive sealing opportunities; unbroken draught-proof finishes.

Resources

Information
Green Building Digest, ACTAC Liverpool
Passive House+ journal, Temple Media, Dublin.
http://shrinkthatfootprint.com/how-much-heating-energy-do-you-use

Less common products
Ancon cavity ties: *www.ancon.co.uk*
www.ecologicalbuildingsystems.com
Hemp-lime: *www.limetechnology.co.uk/hemcrete.htm*
Multi-cavity bricks: *www.wienerberger.co.uk*
Sheepswool, etc.: www.naturalinsulations.co.uk

1. In mild-wintered UK: Burberry, P. (1995) *Green Building Digest 6*, September 1995, ACTAC Liverpool (others say 5%).

2. Roaf, S. (2014) Lecture at Green Architecture Day, Brighton Permaculture Trust.

3. Ibid. Cooling from all four sides is the reason Swedish motorway bridges are warmed.

4. MacKay, D. (2009) *Sustainable Energy – Without the Hot Air*, UIT, Cambridge.

5. It isn't really free, of course: we pay for food, electricity and window glass – but not from our heating budget.

6. This example relates to relatively mild winters. In colder winters, the savings due to incidental heat gains are proportionately smaller, but still significant.

7. In this, my approach differs from most standard recommendations, which consider ground floors to need more insulation than walls. I don't see why.

8. For the British Isles.

9. The thicker the (vapour-permeable) insulation, the deeper in will any condensation occur – so the more important is it to choose a natural, hydroscopic material (e.g. wool, straw, wood-fibre) to prevent a pond forming. Or use a vapour-impermeable insulation material. These, however, are mostly plastic: high embodied pollution and risk of indoor pollution.

10. But note that BRESCU uses 3, 2.5, 5, as conduction is the most effective form of heat transfer – and also possibly because thick underfloor insulation is easy to do, but walls and roofs bring constructional limitations. Damp ground (common in Britain) is a good conductor.

11. °K = (in this context) °C temperature difference.

12. Hall, K. (ed.) (2008) *Green Building Bible*, Vol. 2, Green Building Press, Llandysul.

13. A worry, but not proven.

14. *Green Building Digest 2*, February 1995, ACTAC, Liverpool; *www.greensciencepolicy.org/node/26* (accessed: 29 November 2013); manufacturers and internet information varies).

15. Wienerberger Poroton and Porotherm bricks.

16. Isocyanates are sensitising agents for chemical allergy: Curwell, S., March, C. and Venables, R. (1990) *Buildings and Health: The Rosehaugh Guide*, RIBA Publications, London.

17. Cotterell, J. and Dadeby, A. (2012) *PassivHaus Handbook*, Green Books, Cambridge.

18. Ibid.

CHAPTER SEVEN

Keeping warm: airtightness issues

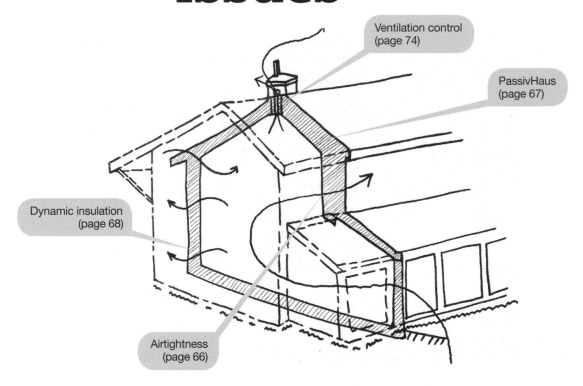

Ventilation control
(page 74)

PassivHaus
(page 67)

Dynamic insulation
(page 68)

Airtightness
(page 66)

Airtightness basics

Warmth makes things expand. Vapours expand greatly, if free to do so. If not, they increase pressure. In houses, this is why the warm indoor air has a higher pressure than cold outdoor air. This causes draughts, and not just draughts: air can percolate through the tiniest of gaps. Just as surprisingly few things are 100% waterproof, surprisingly few are 100% vapour-impermeable. This is both good and bad. Good, because breathing construction assures us of fresh air even when all windows are shut, and bad because, however thoroughly we insulate, lots of heat is lost through ventilation we can't control, and which is subject to the variability of weather. Vapour-impermeable materials with airtight connections can greatly reduce this heat loss – but airtightness is remarkably hard to achieve, particularly as all parts of buildings move (e.g. with drying out, temperature, humidity, vibration and age- and use-related deterioration). Consequently, in well-insulated buildings, most heat loss is through ventilation unless they're airtight and have heat-recovery systems. And, if airtightness is achieved, there's still the issue of air quality to consider.

Neither significant heat loss nor compromised air quality is desirable, but we have to find a balance between them. What the appropriate balance is isn't an absolute but depends on the value you accord to each. Some design approaches (e.g. PassivHaus) uncompromisingly prioritise airtightness. Others (e.g. breathing walls) prioritise air quality. Dynamic insulation combines both objectives by reversing indoor–outdoor pressure differences. However, not only is it hard to do this without mechanical air movement, but also it's vulnerable to reversal by the semi-vacuums that strong winds cause on buildings' lee sides and roof-slopes.

Does this sound complicated? It is. The key principle is simple, though: minimise uncontrollable heat loss – but never to the extent that it compromises health.

PassivHaus

Most recent houses – meeting current energy-saving regulations – use 100kWh/m^2 of heating per year; leaky old ones average 300.[1] PassivHäuser, however, only need 15kWh heating per square metre (10¾ sq ft) per year. (Ironically, however, although small houses need less energy than big ones, smallness increases surface-to-volume ratio, making PassivHaus certification harder to attain, as it's based on energy per unit of floor area.) This minuscule heating demand depends on excellent insulation and, for thermal stability, heavy thermal mass[†]. These are relatively easy to achieve, but on their own can't reduce energy consumption enough. As a result, PassivHäuser depend on exacting airtightness standards: for certification, air changes mustn't exceed 0.6 per hour. This means PassivHaus design relies more on airtight construction than anything else.

To further minimise heat loss, PassivHäuser normally have large south windows, moderate-sized east and west windows and no north ones. (This, however, would produce blank, unwelcoming north façades, leave security blind spots and preclude some rooms from enjoying bi-aspect daylight, so shouldn't be an inflexible rule.) In fact, however, although most PassivHäuser are solar heated, not all are. Indeed, south glazing exceeding that required for daylight only saves £6/m^2/year,[2] so doesn't justify its expense. (This calculation assumes there's no mobile insulation. See: **Chpater 10: Warm windows**, page 105.)

In super-insulated buildings, the only significant heat loss is through ventilation, even if it's carefully controlled. Heat-exchangers[†] can recover half this heat and, by pre-warming inlet air, let us feel comfortable at 2°C (4°F) lower temperature.[3] Besides insulation, therefore, PassivHäuser also rely on mechanically driven ventilation through heat-exchangers (MVHR[†]: mechanical heat-recovery ventilation).[4] MVHR ensures both economy and thermal comfort throughout the year but, like air-conditioning, only works if windows are closed – and the air doesn't *feel* fresh. You can, of course, open windows – but this compromises energy-saving efficiency.

There are two sides to the MVHR versus natural ventilation debate. From an energy conservation perspective, MVHR is crucial for ultra-low energy consumption. It's almost impossible for buildings to reach the British Code for Sustainable Homes Level 6 without it; and it's essential for PassivHaus certification. But for health, it's a concern, and even for energy conservation it's no silver bullet, nor foolproof.

In theory, fresh air should enter houses on the windward side, supply habitable rooms[†], then pass at high level (e.g. over door heads) to moisture-producing rooms, from which it leaves, either to leeward or by ridge vents. This is a simple one-way path. Heat-exchangers, however, collect heat at the *end* of the journey to warm air at the *beginning*. This means ducting. Its layout is critical as overcomplicated ductwork can cause 20% of a building's heat loss.[5] Nonetheless, although this is less than the heat normally lost by natural ventilation, MVHR currently isn't cost-effective.[6] As installation costs are falling and energy costs rising, however, this will probably change.

MVHR's proponents say the fans are inaudible;[7] the air quality better as it's filtered; and the thermal comfort, luxurious. Also that we accept mechanically handled air in cars, so why not in houses?[8] They also point out that natural ventilation is only as good as how it is controlled. Most people, especially when concentrating on other tasks, don't open or close trickle ventilators nor turn on extract fans when they're needed (see: **Airtight construction**, page 70). Against this, however, mechanical ventilation air quality is only as good as filter maintenance. Before choosing this option, ask yourself when you last changed your cooker extract hood filter.

Moreover, from a health perspective, mechanically handled air is questionable.[9] At the very least, it requires careful materials and furnishings specification to minimise toxic emissions (see: **Chapter 13: What we breathe**, page 151). Fan-driven air is also never as invigorating as fresh air. Also, as it's exposed to electromagnetism from the fan motor and friction passing through ducting, it loses most – often all – negative-ions. That it's traditional to open windows at night (or, in cold climates, briefly before going to bed, to flush out old air) isn't just Victorian belief in Spartanism. We breathe more deeply in cool air, so wake more refreshed.

However, there's also a less-studied aspect of the mechanical versus natural ventilation debate: delight. For heating engineers, thermal comfort means constant optimum temperatures. These are sense-dulling and make the body's thermal-regulating mechanisms lazy. *Experientially*, however, thermal delight is a sensation we're *aware of*. Varied thermal conditions stimulate the senses, giving it an aesthetic dimension. Additionally, fresh outdoor air connects us to the ever-changing life of seasons; it nourishes the soul in a way that even-quality mechanically handled air never can (see: **Chapter 14: Nature connection**, page 176). Consequently, despite their excellent contribution to climate protection, I have serious reservations about PassivHäuser. As they cut heating energy consumption by 90%, I wish I didn't. Their meticulous attention to heat conservation, however, is a standard all cool-climate buildings should aspire to.

Keypoints

- PassivHäuser need only 15kWh heating/m^2/year.
- They rely on excellent insulation, airtight construction, high thermal mass, and mechanically driven ventilation air through heat-exchangers.
- They're (usually) solar heated, with large south windows and no north ones.
- They cut energy consumption by 90% but mechanically driven air has health implications.

Choices

Options	Decision-making factors
Airtight or breathing walls?	PassivHaus certification; Level 6 certification, Code for Sustainable Homes; sensory stimulation; air quality; health; attention to operating controls.
Trickle ventilation, dynamic insulation or heat exchangers?	
No north windows or bi- / tri-aspect daylight?	Daylight quality; views; surveillance for security; approachable or hostile façade?

Dynamic insulation

Mechanically driven air through heat-exchangers isn't the only way to recover ventilation heat loss. 'Dynamic insulation' also does this.[10] This system is the complete opposite of PassivHaus airtightness. It works on the principle that if construction – particularly ceilings – is air-permeable and air pressure is lower indoors than outdoors, air percolating inwards through the construction collects heat trying to leave. This also eliminates interstitial condensation risk, as, however high its relative humidity, the incoming air is colder, so contains less water (see: **Chapter 6: Building physics**, page 56). Hence buildings are mould-free and last longer. Air pressure indoors isn't normally lower than outdoors, so this pressure differential must be

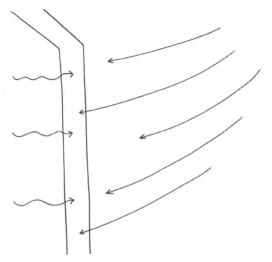

created. It can be induced by stack effect[†], but usually requires extract fans, which can be combined with heat-exchangers.[11]

Wind, however, creates low pressure to leeward, so can actually reverse the direction of the air movement by causing suction. Air, therefore, is best admitted through ceilings, if these are below a plenum[†] space to balance out the (windward) pressure and (leeward) suction on roofs. Also, ceilings usually have a larger surface than walls, so offer less resistance to airflow through them. Perhaps counter-intuitively, walls should therefore be *impermeable* (or nearly so), unless they're very sheltered from winds or rain-screens or a skirt of linked sheds form a pressure-balanced plenum-like space around them.

Dynamic insulation': air percolating in through the construction collects heat trying to leave.

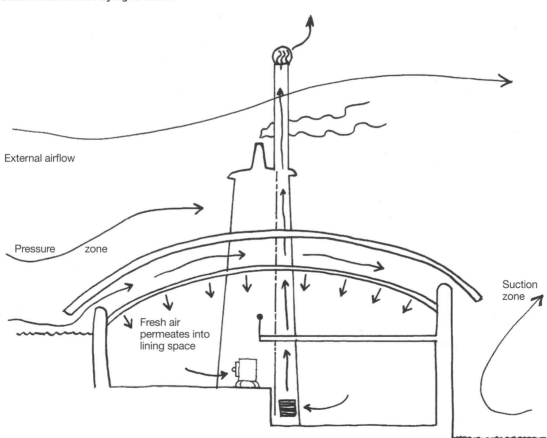

Dynamic insulation: using the ceiling void as a pressure-balancing plenum. Extract airflow is accelerated by warmth from the stove flue.

Unlike PassivHäuser, however, the air *occupants breathe* doesn't have to have passed through the heat-exchanger and associated ductwork and fan. The heat extracted by the heat-exchanger can warm something else (e.g. water or the material that the incoming air passes through), not the air itself. This ensures much healthier indoor air quality.

In theory, a zero dynamic U-value (U_{dyn}) is achievable. In practice, dynamic insulation can halve heat loss in super-insulated buildings using only a third of conventional insulation.[12] In buildings over one storey, however, fans require more power, so use more energy, and air-movement complications are more likely. Dynamic insulation has been shown to work in Scandinavia and Central Europe. Although its energy efficiency has been little tested in milder or windier climates, its indoor air-quality and condensation-prevention benefits are nonetheless indisputable.

The concept of dynamic insulation is simple: it originated from vernacular practice (inlet air through haylofts). To actually work, though, it needs professional calculation, as airflow patterns and speed, materials' porosity and surface area, external wind and temperature, room-to-room transfer and air-leakage rates all affect performance.

Although PassivHaus design and dynamic insulation both require professional input, they're both well within the scope of self-build. PassivHaus design is principally about good practice, and only requires heating/ventilating consultants' input for the MVHR plant and ductwork. Dynamic insulation systems, however, are rather more sophisticated, so require professional design. Both systems require exacting attention to construction detail. Shortcomings in a PassivHaus mean it can't be certified but will still result in a house that is almost as energy efficient. In a dynamically insulated house, on the other hand, shortcomings (e.g. air leakage in the *non*-porous parts) can threaten the whole performance. Nonetheless, this system offers unrivalled potential: greatly reduced (conceivably zero) energy consumption, healthy indoor air and fabric longevity. It may be difficult to calculate but building it is achievable.

Keypoints

- Dynamic insulation is an alternative to MHVR.
- It can reduce heat loss (theoretically to zero), improve indoor air quality and protect building fabric from condensation-related decay.
- It requires comparable levels of airtightness in specific places, to ensure negative pressure indoors, but also air-permeable materials for inlet air to filter through.
- Surfaces for inlet air must be protected from variable air pressures caused by wind (e.g. by using plenums).
- System design is complex, so needs professional expertise.

Choices

Options	Decision-making factors
Extract fans only? Or both inlet and extract fans?	Variable air pressures caused by wind exposure and building shape; stack-effect contribution; ease of creating a ceiling plenum; air quality issues.
Inlet air percolation speed (hence fan capacity and energy)	Surface area and porosity; humidity-buffering capacity; allowance for power or fan failure.

Airtight construction

Airtight construction is essential for all minimum-energy buildings, except those with dynamic insulation. There is, however, no tradition of airtight construction in much of the world, so, when briefing contractors, airtightness needs particular and repeated emphasis. The surest way to make a building airtight is to have one completely airtight layer. If the construction will be by a builder, draw – and highlight – this airtight layer on all plans, sections and details and describe it carefully in construction documents.[13] Also aim for constructional simplicity and easy inspectability. The *PassivHaus Handbook* recommends that all trades meet for an airtightness briefing before starting work, and that one team member is appointed as an 'airtightness champion' to ensure sustained attention to detail.[14] Pressure-testing to check airtightness (and subsequent smoke-tests to locate faults) must be undertaken before you lose access to anything that needs remedying. As this is a timing issue, you need to discuss construction sequence so that inspection and fault-repair are easy but work isn't held up.

To ensure the airtight layer is airtight, it must be unbroken. As all building materials move (e.g. with settlement, temperature, humidity), this necessitates careful attention to detail, especially where different materials

join. You may also need to pre-fix strips of vapour-barriers and wind-barriers between rafters and purlins†. Additionally, as sealing-tape adhesive, sealant foam and mastics have limited lifespans, membrane overlaps should be sandwiched between firm things (e.g. battens and rafters).

Common situations:

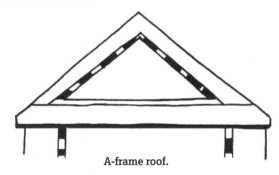

A-frame roof.

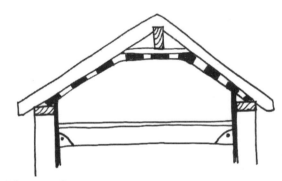

A-frame roof:
Airtightness membrane sealed under wall plaster.
Wall plastered behind joist-hangers.

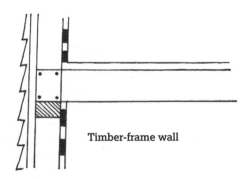

Timber-frame wall

Discontinuity of the vapour-check membrane compromises airtightness (and allows moisture-rich air into the construction, which can lead to interstitial condensation, mould and possibly rot). It's vital, therefore, to ensure membrane continuity at roof-wall and floor-wall junctions.

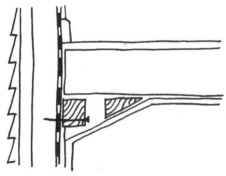

Timber-frame wall:
Airtightness membrane continues without interruption.
Ledge for joists coach-screwed to studs.
Shaped ceiling to conceal (and fire-protect) ledge.

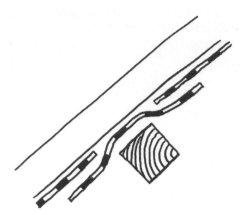

Strips of vapour-barrier between rafters and beams.

For airtight vapour- / wind-barriers, you need firm things between which to sandwich overlap.

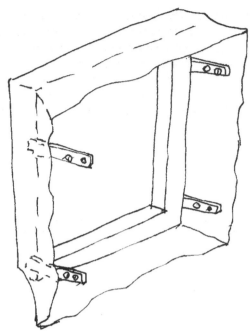

Tape window-frames with windproof membrane before inserting into openings.

Obviously, there should be no penetrations through the airtight layer, but lots of things have to come into buildings through their walls and floors (e.g. pipes and cables). These create major air-leakage risks. Moreover, even if they're well sealed at point of entry, the airspace around cables and pipes connects the air-permeable wall to the holes they come out of. To prevent this, don't chase any services into external walls, but either restrict them to internal walls or surface-mount them. Although surface-mounting can be unsightly, you can hide most in cupboards or under shelves. If penetrations through the airtight layer are unavoidable, seal them carefully.[16] Similarly, in electrical socket boxes, don't bunch several cables through single grommets. (This is no minor concern: I've experienced near-gale-force draughts from electrical sockets on *interior* walls.)

Although draughts come through holes, much air leakage is through masonry. Although its air permeability is an advantage for health (see: **Dynamic insulation**, page 68; **Chapter 9: Making an old building dry**, page 95), for heat conservation, it isn't. External walls therefore need to be airtight. The simplest and cheapest way to seal them is to plaster their indoor faces. Also, to draught-seal gaps where window- and door-frames abut brick, plaster reveals[†] before fitting the frames, tape windproof membrane to the frames and cover with a second layer of plaster.

Air permeability means air can go into external walls but come out of partitions abutting them. You should, therefore, complete plastering external walls before constructing studwork partitions. If partitions are blockwork, plaster the whole lot together. Also plaster walls behind joist-hangers before cutting slots for these; then tape joists, over their hangers, to the walls. This plaster should now form an unbroken airtight layer reaching from ground-floor slab to roof. Before bits get covered up (e.g. by upper floors), check for, and fill, drying-out cracks.

Wet-plastering, however, isn't popular with developers: it's slower than dry-lining and prolongs drying-out time. (Houses built with concrete floors, masonry walls and plaster finishes typically take two years to dry.) Consequently, many recent (or recently renovated) houses aren't plastered inside, but dry-lined. Under this plasterboard[†] is an airway that connects all air-leakage points – making airtightness impossible. (See also: **Chapter 8: Thermal mass**, page 83; **Chapter 12: Keeping cool in an old building**, page 142.) I therefore don't recommend dry-lining and, if it's already there, recommend its removal.

Air leakage cools buildings but draughts cool us. Nobody wants draughts. Nobody wants unnecessary heat loss. This raises a question so politically incorrect that nobody ever mentions it: what if you want to keep a cat? Whatever cat-flap manufacturers say, most will blow open in gales, since gales are stronger than cats. (Freedom Cat Doors, however, claim to be gale-proof.[17] I haven't used these, nor has my cat, so can't comment.) In cold-winter countries (where most PassivHäuser are), cats don't go outside in winter, but have a tray in the hallway. Whatever cat-litter manufacturers say, cat trays smell: not ideal if you have an airtight home. Cat doors behind dense shrubs are probably draught-safe. However, to avoid both unwanted air changes and unpleasant smell, the only way I'm fully confident of is to locate either cat flap or cat toilet in the draught lobby, so it's shut off from the rest of the house. This is unpopular with the cat and, if it requires more space to be built, very expensive.

There remains a yet more serious question: how does anyone living in an airtight house get enough fresh air? In PassivHäuser, it's simple: just rely on MVHR – you don't have to do anything (except pray for no power-cuts). In ordinary houses, building regulations require trickle vents, but who opens them, and how often? To reduce fuel bills, some people never do, and yes: they're still alive – just. (*Under*-ventilation, however, is more a problem with those who have had airtight construction imposed on them than with homeowners who have chosen it.[18]) Indoor air can contain water vapour, carbon monoxide (CO), CO_2, and indoor pollution from furnishings and construction materials (see: **Chapter 9: Keeping dry**, page 91; **Chapter 13: What we breathe**, page 151). Concentrations of these can reach levels injurious to health. Moreover, like all animals, we need air to live. But we also want to be warm. How can we reconcile minimum ventilation to minimise heat loss with ventilation adequacy for health? As described in the next section, we need to link airtightness to a strategy for controlling ventilation, and we need foolproof methods of achieving this.

Keypoints

- Masonry is air-permeable unless plastered.
- Design an airtight layer. Highlight this on all drawings and documents.
- Aim for constructional simplicity and for easy inspectability.
- Hold an airtightness briefing for all trades before starting work.
- Appoint an 'airtightness champion'.
- Pre-fix strips of airtight membrane in inaccessible places (e.g. between rafters and purlins) and sandwich membrane overlaps (e.g. between battens and rafters/studs).
- Minimise the number of places where the airtight layer is penetrated and seal all penetrations, including cables entering electrical socket-boxes.
- Wet-plaster the indoor faces of all external walls from ground-floor slab to roof in an unbroken layer, including external walls behind joist-hangers and over the sealing tape around window- and door-frames.
- Before covering anything up, fill drying-out cracks.
- Pressure-test on closure of the external envelope, installation of all services and completion.
- Most importantly: decide a ventilation strategy – and methods of achieving it.
- And don't compromise airtightness with cat flaps.

Choices	
Options	**Decision-making factors**
Breathability or airtightness?	Attitude to fresh or mechanically handled air; health issues; dynamic insulation or PassivHaus airtightness; opportunities for controllable ventilation.

As airtightness is an unfamiliar concept, to achieve it, there's no alternative to meticulous attention and adherence to detail.

Ventilation control

Well-insulated buildings lose most heat through ventilation. We need air, of course, but not uncontrolled ventilation: draughts. Controlled ventilation depends on draught-proof *construction*, as well as draught-proof windows and doors. But we mustn't exclude all air. Without enough air, open fireplaces (exceedingly inefficient, so not recommended) won't burn well and stoves will produce odourless carbon monoxide (CO), which could kill us. Due to under-ventilation, chronic low-level CO poisoning is surprisingly prevalent. (Low-level CO is common in kitchens, as cooking with pans much larger than gas rings restricts oxygen supply to the flame; and closed windows and extract fans turned off keep the CO in the room.[19]) At the very least, CO_2 will make us drowsy; and humidity will grow mould and breed dust mites.

We may need fresh air, but in midwinter, it's bitter. Inlet air, however, can be 'temperature-tempered'. It can be solar-warmed (e.g. by air-heating panels), even a little on overcast days, but the warmth won't extend long after dusk. If we place inlet ducts at a low level behind radiators, they allow air in without letting heat escape. Being warmed by the radiator, the incoming air will rise into the room. And – since warm air rises – no air can drain out through the duct. Underfloor (and under-insulation) inlet pipes allow in air that is the same temperature as the surrounding earth, i.e. close to the annual average temperature (in lowland Britain, mostly around 11°C/52°F[20]). If these inlet pipes rise behind the stove (and are perhaps also warmed by

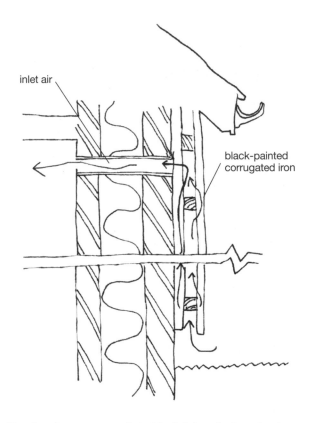

inlet air

black-painted corrugated iron

Simple solar pre-warmed air. The brickwork stores heat, but as it's at a low temperature, not for long.

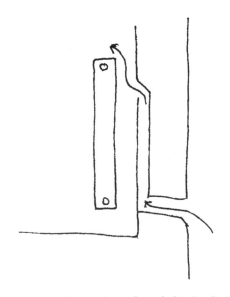

Pre-warmed inlet air: rising ducts behind radiators.

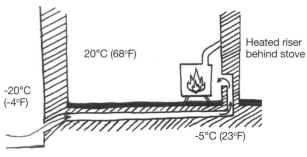

20°C (68°F)

Heated riser
behind stove

-20°C
(-4°F)

-5°C (23°F)

Underfloor inlet air.

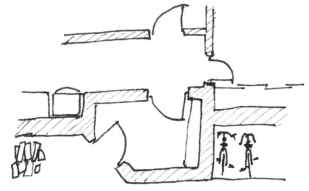

Draught lobby.

pipes to radiators) they will only permit one-way air-flow, so that heat can't escape (for outlet ventilation, see: **Chapter 9: Keeping dry**, page 91).

As lots of heat can be lost when opening outer doors, entrances should, if possible, be on the sheltered side of the house or sheltered from prevailing and cold winds by shrubs or fences. Even better than this, is to have a draught lobby. Although cool air from the lobby will enter the house when the connecting door is opened, its amount is small, and it's slightly pre-warmed by being semi-indoors. Consequently, a draught lobby greatly reduces heat loss from unintended air change. It's also a good place to hang rain-soaked coats, so they don't make indoor air humid. Even if the lobby's external wall is insulated, because it sometimes fills with cold air you need further insulation between it and the rest of the house – just as for larders. I therefore usually put half the thickness of insulation that other external walls require in the house-lobby wall, and half in the lobby-outdoors wall.

Notwithstanding all such draught-control measures, however, we can't do without air change. Besides air inlets dedicated to combustion appliances, we ourselves need fresh air. In summer, we crave it. In both winter and summer, we also need plenty of ventilation options to accommodate the vagaries of weather (e.g. wind direction, driving rain, blizzard-eddies bringing snow indoors, dead calm) and cover the full range from trickle to full flush.

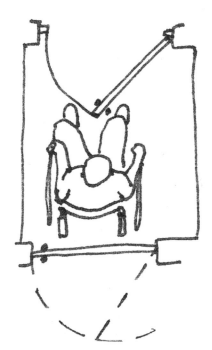

Stuck! Draught lobbies must be large enough for wheelchair and pushchair access.

Keypoints

- Eliminate draughts and air leakage through building fabric by careful construction detailing.
- All combustion appliances need fresh-air inlets.
- Provide opportunities for trickle ventilation.
- Provide plenty of ventilation options to accommodate the vagaries of weather.
- 'Temperature-temper' inlet air in winter.
- Provide a draught lobby at the main entrance.

Choices

Options	Decision-making factors
Heat exchanger, extractor-fan, trickle vents or temperature-tempered inlet air?	Energy conservation; air quality; carbon monoxide risk if there's a stove or gas cooking; humidity sources.
Draught lobby or minimum footprint?	Planning regulations on façade extension; compact insulation perimeter; draught lobby inside or outside insulation perimeter.

Resources

Information

Burberry, P. (1978) *Building for Energy Conservation*, Wiley, London.

Cotterell, J. and Dadeby, A. (2012) *PassivHaus Handbook*, Green Books, Cambridge.

Passive House+ journal, Temple Media, Dublin.

Less common products

Airtightness tapes: *www.proctergroup.com*

Daylighting, ventilation and cooling: *www.monodraught.com*

Freedom Cat Doors: *http://energyefficientdogdoors.com*

Intelligent membranes: *www.imaroofer.com*

1. UK figures: *http://shrinkthatfootprint.com/how-much-heating-energy-do-you-use* (accessed: 22 May 2014).
2. As installation costs 80p/kWh (in UK), some consider more south glazing than necessary for daylight isn't justifed: Nick Grant, quoted in de Selincourt, K. UK PassivHaus Conference 2013 review, *Passive House+*, 2013, Issue 5. This calculation subtracts heat losses from solar gains. Were insulated shutters fitted – and used – the figures improve significantly.
3. Cotterell, J. and Dadeby, A. (2012) *PassivHaus Handbook*, Green Books, Cambridge.
4. Similarly, heat-exchangers can use shower drains' 35°C (95°F) average water temperature to pre-heat incoming water to 25°C (77°F): *www.renewability.com/power_pipe/index.html* (accessed: 18 March 2015). But as solar can do this even in winter, it's an alternative, not addition, to solar heating.
5. de Selincourt, UK PassivHaus Conference 2013 review.
6. Feist, W. (2013) Interview, *Passive House+*, 2013, Issue 5. Energy-wise, however, it is.
7. At 25 dBA (inlet) and 30 dBA (outlet in kitchen or bathroom/toilet), the PassivHaus Institute conditions for certification, it's just audible. (If it's louder, you won't get certification, but nor will you demolish the house.) So acceptability depends on how much noise you're used to. However, background noise, so low level that you're unaware of it, and complete silence aren't the same. The difference is stress; and the relief you feel when it stops.
8. Personally, I would rather have a window open, but I may be alone in this.
9. Worse, some people use plastic drainpipe as ducting. This is smooth-bore and cheap – but it's PVC, so off-gasses toxic vinyl-chloride, especially when warm.
10. Halliday, S. Performance of dynamic insulation – Gaia: *www.gaiagroup.org/Research/techinv/DI/cibse_di.pdf* (accessed: 18 September 2014); *www.sciencedirect.com/science/article/pii/S2212609013000046* (accessed: 16 September 2014).
11. Ibid.
12. *Building Magazine*, 13 February 2009.
13. Ibid.
14. Cotterell and Dadeby (2012) *Passivhaus*.
15. Ibid.
16. McGuiness, S. (2014) Airtightness: the sleeping giant of energy efficiency, *Passive House+*, 2014, Issue 7, Temple Media Ltd, Dublin.
17. *http://energyefficientdogdoors.com*
18. de Selincourt, K. (2014) Natural ventilation: does it work? Passive House+, 2013, Issue 6.
19. de Selincourt, K. (2014) Healthy buildings must be warm, well ventilated and dry, *Green Building* magazine, Spring 2014, Vol. 23, No. 4.
20. MacKay, D. (2009) *Sustainable Energy – Without the Hot Air*, UIT, Cambridge; although air-temperatures average 12-14°C (54-57°F): *http://www.metoffice.gov.uk/climate/uk/averages/ukmapavge.html* (accessed: 19 September 2014).

Keeping warm: heating

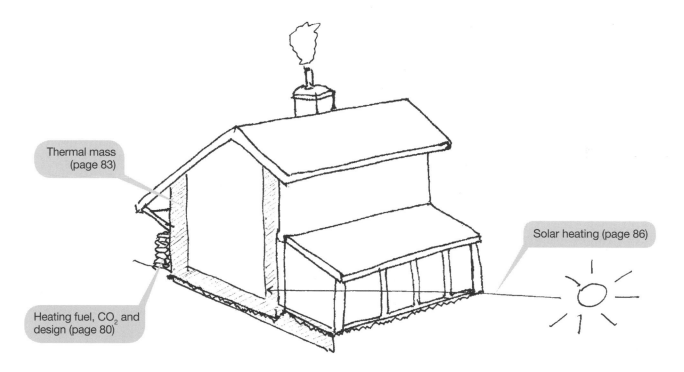

CHAPTER EIGHT

Thermal mass (page 83)

Solar heating (page 86)

Heating fuel, CO_2 and design (page 80)

Heating basics

Heating buildings *isn't* about heating us. It's about slowing the rate at which our bodies cool. We need our bodies to cool because metabolism produces heat – and if we don't lose some of this we'll overheat and die. However, if we cool too much, we'll get hypothermia: also fatal. Heating buildings, therefore, is about letting our bodies cool *at the right rate*: that's why, although we may sometimes absorb radiant heat, we more often just need to slow the rate at which *we* radiate heat, and want the air in our homes at neither sauna nor cold-store temperatures. We, like buildings, cool by conduction, convection and radiation. How warm or cold we feel is always due to a combination of these. Consequently, if we lose less heat by one means we need to lose more through the others to keep cooling at the right rate. For example, if we warm the *air* in a room, the rate at which we lose heat by convection will slow and we'll feel warmer. Or, if the surfaces around us are warm enough to reduce the heat we radiate to them, we need cooler air for comfort.

Keypoints

- Space 'heating' is actually about *slowing* our body cooling rate.
- Thermal comfort depends on a combination of conduction, convection and radiation. More of one means we need less of the others.

Heating

Air temperature is easy to measure; radiant heat isn't. As a result, heating standards are normally based on air temperature. But which is better: warm air or radiant heating?

Forced air heating is the cheapest system to install, the fastest way to warm rooms and isn't damaged by freezing, but, for health, it's the least desirable. Heated air carries dust – carbonised by heaters – which irritates throat and lungs.[1] Even convection from radiator surfaces above 45°C (113°F) does this. (Radiators are misnamed: they're designed to primarily heat air, so heat much more by convection than radiation.) Heated air also tends to stratify thermally, leading to 'hot head, cold feet syndrome'; or worse, hot upstairs bedrooms (when nobody is in them) and cold downstairs rooms (so we turn up the heat). Additionally, the warmer the air is inside a building, the more heat is lost through ventilation. (Also, if warm air rises into upstairs bedrooms and is then ventilated away, lots of heat is lost here.)

Radiant heat avoids these disadvantages and also warms the body more deeply. However, radiation declines with distance, and heat, just like light, radiates in straight lines, so anywhere out of sight from the heat source is shaded from primary heat. (This isn't necessarily a disadvantage: focal warmth is socially focusing, so people tend to gather round stoves.) Just like sunlight, however, radiant heating warms other things and surfaces, which both warm the air and radiate to us, warming us with 'secondary' heat. Radiation delivers warmth – or slows radiation from our bodies – most effectively if the warmth-radiating surfaces are all around the room (i.e. walls and/or floors), so some of it is direct and strong radiation can balance weak.

'Radiant' floors – underfloor heating – are easy to install. Usually, for even heat distribution, pipes are covered with a few inches of thermal mass (e.g. concrete, lime-mortar). Consequently, they deliver heat slowly and steadily; a very stable climate but one that can't be rapidly adjusted to other heat inputs (e.g.

Radiant floors: large radiant and convection surface, but little body exposure.

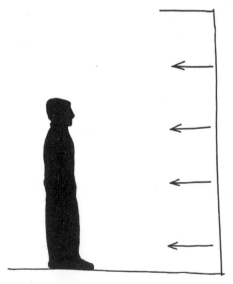

Radiant walls: maximum body exposure, but easily obscured radiant surface.

solar, occupants, weather change; but see: **Thermal mass**, page 83). Underfloor heating radiates to the whole room and heats air without stratifying air temperature or propelling airborne dust. Much of their heat transfer, however, is conductive: through the soles of our feet.

Radiant walls are less common than radiant floors but give better warmth exposure to our whole body. (Fixing anything to walls, however, needs especial care: use pipe-finder instruments or turn the heating on, then spray walls with a spray bottle so evaporation patterns disclose the pipes' locations.) Wall panels are manufactured, or you can build in piping between two leaves of a brick/block wall. (Allow access to fittings for maintenance, even if piping is non-corrodible plastic.) Best, of course – but most expensive – is to have both radiant floors *and* walls. With either radiant walls or floors, to compensate for obstruction by rugs, furniture, shelves and suchlike, the warmed area normally needs to be slightly larger or warmer than that optimally required.

Skirting heaters all round a room are cheaper than underfloor heating. These can run at lower temperatures than radiators, so give mostly radiant heat and no convection-propelled dust. They also warm the wall surfaces above them (especially if these aren't electrostatic, e.g. painted with acrylic), which also radiate. Some (e.g. ThermaSkirt) are only 20mm (¾") thick, so can replace skirting boards instead of taking up space. This system, however, limits the amount of furniture you can have around walls, otherwise heat won't radiate directly to occupants.

Unlike forced or convected air heat, which warms the skin, all-round radiation (especially from radiant walls *and* floors) so deeply warms the body that you only need 19°C (66°F) to feel as comfortable as in 22°C (72°F) air. This 3°C (6°F) reduction means more (and delicious) comfort for significantly less energy.

Keypoints

- Forced air heating and radiators over 45°C (113°F) circulate carbonised dust.
- The warmer the air, the more heat is lost through ventilation.
- Radiant heating is limited to line-of-sight and declines with distance – but warms other things, which then radiate to us.
- Radiant floors are large surface area, but much heat transfer is conductive, to our feet.
- Radiant walls radiate to the whole body but are easily partly obstructed by furniture. They necessitate care when fixing anything to walls.
- Skirting heaters all round rooms are cheaper than underfloor heating and give mostly radiant heat from low temperature surfaces. They also are easily partially obstructed by furniture.
- All-round radiation reduces comfort air temperature by 3°C (6°F).

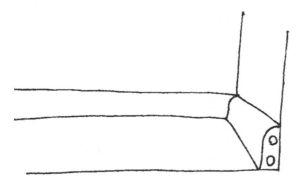

Skirting heaters: all-round radiation, but easily obscured by furniture.

Choices

Options	Decision-making factors
Forced air, radiators, stove(s), radiant walls, underfloor heating or skirting radiators?	Air quality and health implications; speed of response; thermal stability/ mass; social focus; varied temperature zones or uniform climate; cost.
What balance between conduction, convection and radiation is appropriate?	Comfort; thermal economy; thermal stimulation; ventilation rate; health concerns.

What fuel to use: CO_2 and design implications

What is the best fuel to use? Obviously, it would be best to use no fuel at all, just solar heat (see: **Solar heating**, page 86; **Chapter 18: Solar energy**, page 208). However, in most places there's too much gloomy winter weather to be able to do without at least *some* back-up heating (or, expensively, inter-seasonal heat storage).

Of fossil fuels, natural gas produces the least CO_2. Gas needs professional installation and regular servicing as faulty burners can produce carbon monoxide or cause explosions. When buying a gas boiler, get the most efficient condensing boiler you can. It's also important to choose one that will accept high-temperature, solar-heated input water. Many makes don't. If you already have a boiler but it's pre-2005, you'll probably need to add a flue gas heat-recovery unit.[2] Most post-2005 gas- and post-2006 oil-boilers are condensing types.[3] Oil produces more CO_2 than gas. It can also produce carbon monoxide (CO), but, unlike with gas, it gives you some warning through the foul smelling sooty smoke produced. Coal produces the most soot, smoke and CO_2; if deprived of air, it too produces carbon monoxide.

In airtight buildings, CO poisoning – even from cookers – becomes an increased risk, making CO alarms even more important. Does this encourage you to choose electricity? Electricity is completely fume-free and clean *in use*, but this conceals its massive climate impact. Since most electricity is produced by burning fossil fuels, and since the generation and distribution process is typically only one-third efficient, electricity is the most climate damaging of all forms of energy.[4]

Heating fuels: Kg CO2/kWh[5] There is at least a fourteen-fold variation in climate impacts.

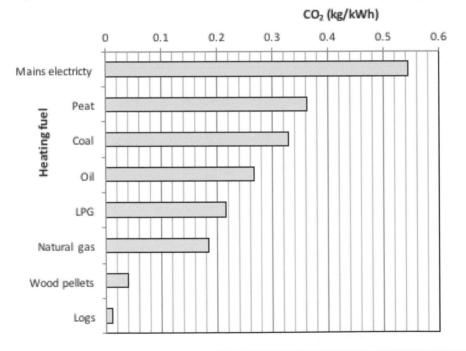

Coal	Oil	Gas	Wood (log, scrap, woodchip, pellet)	Electricity
Produces mostly CO_2, some water.	Produces some CO_2, some water.	Produces mostly water, some CO_2.	Only produces CO_2 captured when it was growing, some water.	Conventional generation produces lots of CO_2.

Burning wood only produces the CO_2 that the wood absorbed by growing. (Without enough air during combustion, however, it too produces CO.) Wood stoves are therefore CO_2-neutral, but they produce smoke. This may smell nice, but it's harmful to health. In towns, therefore, it's essential to use stoves that burn at high temperatures. This ensures gasses are fully burnt, which (virtually) eliminates smoke toxins. At 900-1,000°C (1,650-1,830°F), combustion is also some 8 times more efficient: more heat, lower bills, less smoke. If you connect a stove to an old chimney, this will probably need to be re-lined to prevent leaks. Smoke leaks from open log fires smell cosy; but from closed stoves, leaks bring odourless but lethal carbon monoxide. Alternatively, retrofit a modular-block or double-walled stainless-steel chimney; or build a new brick one with clay flue-liners (BS EN 1457: 1999). Also, the room with the stove in must have a fresh-air inlet.[6] To prevent you from having to sit in a draught, it's best to pipe the fresh air to near or behind the stove, or fit the air-inlet pipe directly to the stove itself.

The other safety risk is fire. All exposed flue-pipes should be cast iron or double-walled stainless steel, separated from combustible material at least as much as regulations require: better is safer. Single-walled sheet steel, common in America, can easily burn through, so is dangerous. In gale-swept places, a draught stabiliser in the flue can keep the fire under control by admitting air above the firebox.[7] Without one (or a post-combustion air inlet on the stove), you risk a chimney fire, house fire or, at least, a cracked chimney. To further minimise fire risk, wood stoves should be appropriately sized. When slow burning, they don't fully combust fuel, so combustible creosote condenses in the chimney (which we tend not to sweep as often as we should). Consequently, oversized ones are potential fire risks.[8] As stoves burn most efficiently at full output, they're therefore best sized for 60-70% of peak load.[9] (On the rare occasions that their full output is inadequate, just move closer to the stove. This minor inconvenience is preferable to burning the house down.)

If, of course, you only light a fire on Christmas Day, it's not worth buying a stove: just use an open fireplace. But if you use one for heating, it definitely is. Stoves are eight times more efficient than open fireplaces. Those with glass doors let you see flames. Opening the doors on the few occasions we sit around them lets us also smell and hear the fire for only brief sacrifices of efficiency.

Wood stoves are primarily a source of radiant heat, but they also heat air (and some have double walls to turn them into convectors), and, with back-boilers, can heat water. The lower they're placed, the better they warm room air – and our legs and feet – but the more can furniture obstruct direct radiation to anyone not close to them, and the lower you must bend to cook on them. Stoves with back-boilers heat water, which can be used for radiators in other rooms. However, if you choose this whole-house-heating function for the stove, be aware that this can cool the stove down so much that it's no longer useful for radiant, socially focusing, room heating.

Stoves designed specifically for water-heating needn't burn all the time as heated water can be stored in large accumulator tanks[†]. Tall tanks store water in temperature-stratified layers. This enables multiple inputs – even low-temperature ones (e.g. lukewarm solar-warmed water in winter) – to feed the tank, and different uses to be fed different temperature water. If the accumulator is in an insulated cupboard, just opening the cupboard doors makes the house warmer.

Instead of storing heat in water in accumulator tanks, some kinds of stove have sufficient thermal mass to keep warm overnight. These include soapstone-faced stoves and Kachelofens[†]. Of all stones, soapstone retains heat longest, much longer than iron. Kachelofens burn small amounts of wood very fast (so correspondingly hot) achieving 70-90% efficiency. Their long flue paths and thermally massive construction lets them store this heat so they can emit warmth all day from a single firing.[10]

Wood stoves can be fuelled by logs, carpentry waste (if it doesn't contain preservatives or toxins) or dry garden prunings. Some kinds burn wood pellets[11] and can be automated, so require little attention. Wood pellets, however, must be stored under cover, and aren't everywhere available.

As moist wood burns poorly and deposits creosote, firewood should be dry and well seasoned. This means you need somewhere to season and store it: an overhanging roof, pole barn or well-ventilated shed. As firewood comes with a resident wood-eating beetle population, it's not wise to store more than a few day's supply indoors.

Stoves will also burn peat ('turf' in Ireland). Its smell is central to peat-burning cultures. However, although classified as a 'slow-renewable fuel', peat actually releases 10% *more* CO_2 equivalent (mostly methane) than coal: mostly during drying prior to burning. Consequently, for mood or atmosphere, it's best burnt only to scent rooms, then followed by wood – if available – or gas-fuelled heating.

Wood stove heating affects house design. The hearth is the domestic 'heart'. It's socially focal; both its protective mood and the warmth it produces are best at the centre of a home. Additionally, chimneys draw best – and leak least – on the roof ridge. At the centre of the house, they warm the rooms around them.

(Mine warms four rooms, besides the one with the stove in.) On gable walls, they just warm outdoor air. For heat distribution, stoves, and the warmed walls behind them, heat the air around them, which rises and strokes ceilings, so exposed beams can trap warm air or direct it to where you want. Wood stove-heated houses, therefore, are best with compact plans but tall: two (or more) storeys. This is the opposite of houses in windy locations where ground-hugging forms suffer the least wind-chill.

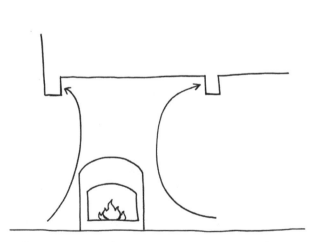

Ceilings and beams form inverted rivers and lakes of warm air. Design can use this for distributing or localising heat.

Least heat loss from wind.

Best heat distribution.

Choices

Options		Decision-making factors
Gas		Availability; suitable tank locations (if no mains supply); low CO_2; low smoke emission requirements (e.g. smokeless zones).
Oil		Convenience and gravity-fed storage.
Coal		Effort; easy storage; dust; legality (smokeless zone?). Is CO_2 acceptable?
Peat		Local tradition (especially in Ireland) but high CO_2 equivalent. Can peat burning be limited to brief periods?
Wood	Logs and wood waste	Availability and time/effort to manage; legality (smokeless zone? high-temperature stove?); storage.
	Wood pellets	Availability and storage.
Electricity		Price; convenience; source (fossil fuel, nuclear or renewable).
Wood stoves: high or low level?		Occupants' proximity to stove (low for a small room, tall legs for a large space); cooking top?

Thermal mass

Indoor comfort depends on thermal stability. Moreover, comfort temperatures are a narrow range: about 4°C (8°F), although up to 6°C (12°F) can be tolerated.[12] However, outdoor temperatures vary widely and can change fast. Even on bitter days, the sun radiates heat – but one cloud and it's shut off. Occupants add some 100w/person (more if they're active) – but they come and go and activity level varies. Electric and electronic devices add yet more heat[13] – but their type, number and use can't be anticipated. Cooking produces heat. Outdoor air temperatures also change, and so, with rain and wind-chill, do cooling rates. With such unavoidable ever-changing heat inputs and losses, temperatures easily fluctuate between too cool and too warm.

Despite this, a building's job is to provide comfortable temperatures. The best way to do this varies according to the building's use. Intermittently occupied, hence intermittently heated, buildings (e.g. churches, meeting halls) cool down between occupancy periods so need a quick response to heating whenever they're used. Conventionally, therefore, such buildings heat the occupants, not the air, with radiant heaters (at the price of comfort, as they're usually positioned overhead: warm heads, cold feet – the exact opposite of what we need). Warm, moist air (from people breathing) and cold walls, however, maximise condensation risk. Insulation on interior surfaces both prevents surface condensation and lets rooms warm up rapidly. Although it obstructs thermal storage, allowing temperatures to fluctuate rapidly, for rarely occupied buildings, this is the most energy-conservative strategy.

In the era before good insulation, it used to be argued that commuter homes need rapid thermal response so they can be unheated when occupants are out at work. The contemporary version of this is using smartphone apps to turn heating on before you get home. This (arguably) *might* save heating energy but does nothing to prevent summer overheating (as smartphones can't control solar-heat inputs): an increasingly serious problem. App-controlled electrically operated sun shades will help, but heat also radiates from the ground and surroundings, and shades can't moderate air temperatures. This approach, however, assumes a reliable electricity supply – which isn't predicted to be the case. Moreover, it ignores the need for 'homeliness'. This depends on 'mothering' stable warmth.

Buildings that are frequently occupied require a different approach: these need thermal comfort – which means stable temperatures. Thermal stability without constant energy inputs requires lots of thermal mass: material (e.g. masonry, cob, rammed earth, concrete) that warms up and cools down slowly, and therefore stores heat. All thermal mass contributes something, but it's normally only worth deliberately building enough for 24-hour thermal stability, for example 100mm (4") dense brick walls, 200mm (8") if exposed to rooms on both sides. Thicker walls store more heat (brick kilns, for instance, stay warm for months), but as the heat emerges slowly, it's hard to perceive their impact on thermal comfort, unless they're very warm (like night-storage heaters).[14] Thermal mass will absorb more heat if it is directly warmed by a radiant heat source (e.g. sun, stove chimney), rather than just

Where best to locate thermal mass

For solar-heat storage	Where sunlight falls, so it's warmer than room air.
For temperature stability	Anywhere in occupants' view (for radiant heating/cooling) and/or stroked by air currents (for convective heating/cooling). Obscured or out of the way thermal mass has only limited effect.
For warmth retention in cold climates	Floors (as warm air rises) and walls (as warmth radiates).
For 'coolth' retention in warm dry climates	Ceilings (as cool air drops) and walls (as we radiate body warmth to them). But also floors (as, although these cool *us* less, cool night air cools *these* most).

For thermal stability, thermal mass should be inside the insulation.

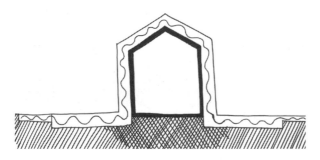

An insulation skirt can let dry subsoil significantly contribute to inter-seasonal thermal stability. (Wet subsoil conducts heat away; and flowing groundwater washes it away.)

being warmed by room air. It warms us most effectively if it radiates directly to us (i.e. by line-of-sight) and/or is stroked by room air. Out of the way or obscured by wall coverings, plasterboard (sheetrock) dry-lining, or by furniture, it's only of limited use. To be any use, thermal mass must, of course, be *inside* the insulation.

There are a variety of materials that can be effective as thermal mass. Heavy material is the simplest and normally cheapest and least environmentally damaging. The most common materials are brick, concrete, stone or rammed earth, which also perform a structural function. You can use anything from soft earth and unfired clay bricks to scrap metal or rubble in gabions[†], though. Large tanks of water store heat well. All thermal storage involving liquids, however, brings containment complications and/or expense.

Some thermal mass materials aren't heavy but use phase-change: like ice, which melts into or solidifies from water, phase-change materials[†] (e.g. styrene wax, eutectic salts[†15]) store heat (or coolth[†]) in latent form and release it when they reach their solidifying (or melting) temperature. This has the advantage that heat or coolth can be stored at close to room tem-

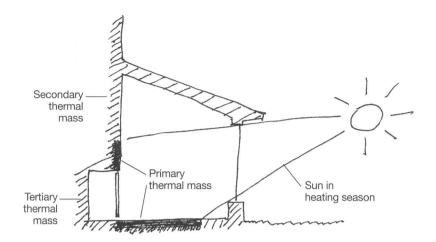

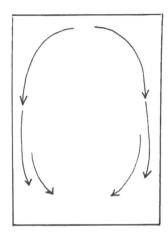

Primary thermal mass stores direct solar heat, which is warmer than air temperature. Secondary thermal mass contributes thermal stability by slowing temperature change. Tertiary thermal mass contributes only a little to thermal stability, as it's not stroked by room air.

Water stores heat efficiently as convection currents expose only its cooler outer layers to cooling.

perature, so, due to the small temperature differential, there's very little loss. These materials can be tailored to change phase between liquid and solid once room temperatures start to drop below (or, for cooling in hot climates, rise above) desired levels. The simplest of these materials is phase-change plasterboard. Although lightweight, this has high thermal capacity. Its core, however, is petroleum-based styrene, so embodies oil-industry pollution.

Masonry buildings are built of heat- / coolth-storing materials, but wood-frame ones aren't, so can cool down or warm up too fast for comfort. How can thermal mass be incorporated into timber-frame buildings? Options include: expensive phase-change plasterboard; thermally retentive solid (e.g. brick, clay) partitions and interior linings to exterior walls; and, if possible, solid floors. All these, however, bring disadvantages: phase-change plasterboard is ecologically questionable; solid partitions complicate remodelling; solid floors are harder on the back than wooden ones; and in seismic areas, it's dangerous to have heavyweight construction above chest height, unless its stability is guaranteed (e.g. reinforced concrete). However, if you can't use any of these, double plasterboard (totalling 1"/25mm) or 25mm clay plaster on walls and ceilings can significantly dampen diurnal thermal extremes. Nineteenth-century lath and plas-

ter performs similarly. As with most things, there's no one best way. It depends on the circumstances – and your priorities.

One way or another, however, buildings need thermal mass. Except in hot humid climates – in which air is barely cooler at night than in daytime – thermal comfort depends on thermal mass for thermal stability.

Keypoints

- Thermal mass stabilises temperatures and ensures warmth/coolth if power-supply fails.
- To contribute to thermal stability, thermal mass should be *inside* the insulation.
- For thermal stability, maximise thermal mass, especially where it radiates to occupants and/or is stroked by room air – 100mm (4") solid walls give 24-hour thermal stability; 200mm (8"), if exposed both sides.
- For heat storage, thermal mass should be directly warmed (e.g. behind a stove, or in sunlight).
- For coolth storage, thermal mass should be where it can be most cooled, or – to cool us – overhead.
- The simplest thermal mass is heavyweight material. But, in seismic regions, remember safety issues.

Choices

Options	Decision-making factors
Insulation outside or inside thermal mass?	Stable indoor climate or rapid response; frequency and density of occupancy.
Locate thermal mass high, low or in sunlight?	For cooling, heat retention or solar heating? Seismicity and material; structural considerations.
Masonry, water or phase-change thermal mass?	Space; reliability; environmental impact; cost; earthquake risk.

Solar heating

The sun provides free, non-polluting heat. Solar heating works on the 'greenhouse' principle: more (long-wave) sunlight passes in through glass than (short-wave) heat passes out through it. Nonetheless, when it isn't sunny, even the best glazing lets a lot of heat escape.

Solar heating is only needed in winter, when sun angle is low, but is undesirable in summer, with high sun. This means south-facing glass *walls* work well. Conversely, glass *roofs* freeze occupants in winter and boil them in summer. (Solar control glass somewhat mitigates summer overheating – when you could be outside – but denies solar heat in winter – when you can't.) Glass roofs give heat in shoulder seasons and lots of light, but summer and winter disadvantages usually outweigh these benefits. Multi-layer polycarbonate roofing also gives lots of light and is insulating – but not as insulating as an insulated roof.

Sunlight varies, so temperatures can fluctuate widely. Consequently, better than glass-walled living rooms are dedicated sunspaces, winter-gardens and conservatories. These are more tolerant of temperature swings than house interiors, so can buffer temperatures before they reach indoor rooms. They can be along the house's front, back, side or end walls. (At the front, however, they're less private, although sloping glazing reflects more sky, partially offsetting this. Sloping glazing, however, is harder to fit shutters to.)

Glass roofs freeze in winter, fry in summer.

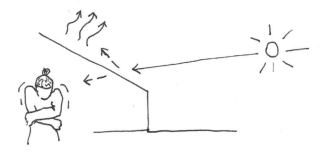

During summer

During winter

Glass walls admit winter but not summer sun.

Slightly inclining glass slightly increases heat collection – and (because it reflects more sky) privacy.

They can be single-storey or several-storey – as long as you can reach, or mechanically open, windows to ventilate upstairs rooms. Heat from these spaces can be transferred to indoor rooms by opening doors or vents. Like sun-traps outdoors, sun-rooms can be in protected corners, and are best focused on noon sun: even in March, there's little worthwhile solar heat before 9.30am. For midwinter heat, they should face within two hours (approximately 30°) of south; for shoulder seasons, within 45°. The further they are out-side these ranges, the less heat they'll get in winter (when you most want it), although they'll still warm up from late spring to early autumn.

Sun-rooms should have *no other heating* as, with so much glass, they cool fast at night and in cloudy weather, so any heat you add to them is quickly lost. (Indeed, heated conservatories waste more heat than they gain by solar heat; far from being heat collectors, they're heat *wasters*.) Conservatories, and all solar-heated rooms, need lots of thermal mass (in both walls and floor) to moderate temperature swings. As thermal mass stores heat without undue temperature rise, its capacity to keep surfaces relatively cool can increase heat collection efficiency by almost 50%.[16] To absorb solar heat, which carries much more warmth than air, thermal mass must be where sunlight falls on it. To maximise absorption of solar radiation,

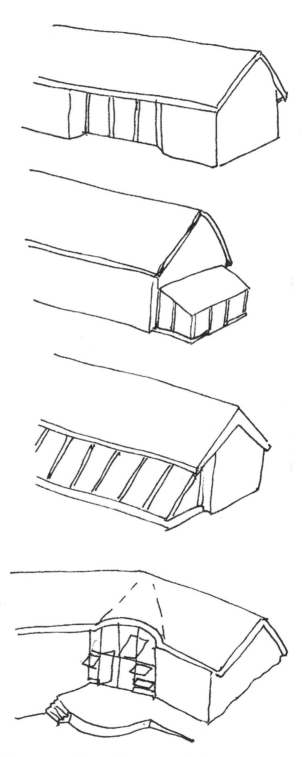

Sunspaces, winter-gardens, conservatories.

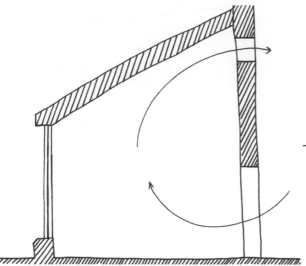

Heat transfer into rooms by doors or vents.

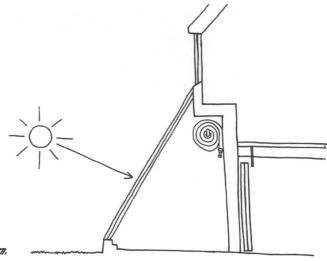

Glass doors between living room and sun-room. Insulated blinds to keep both thermal-store wall and sun-room warm at night.

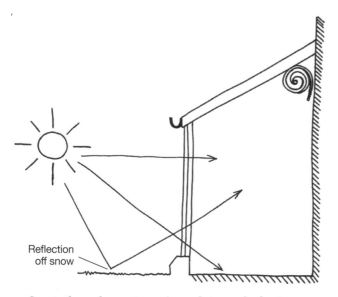

Reflection off snow

Locate thermal mass to receive and store solar heat.

Sun-rooms off houses that don't face south can still maximise south-facing glazing.

Solar space heating[†] depends on seven basic principles:

- High insulation levels, including draught-proofing.
- Large windows to the south, modest-sized ones to east and west, small (or none) to north.
- Good insulation at night time (and in dull, bitter weather) – from multi-layer curtains to insulated shutters.
- High thermal mass – e.g. water, phase-change material or heavy masonry / earthen material (the last is simplest, cheapest and least likely to have problems, although slightly less effective at thermal stabilisation).
- Buffering spaces – e.g. conservatories.
- Overheating prevention – e.g. south-west / west shading in summer, cross-ventilation, solar chimney with cool-air inlet.
- Simple, intuitive operation.

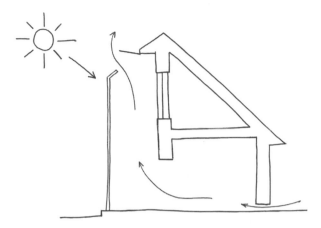

Sun-room as solar chimney for cooling.

its surface should be moderately dark-toned (e.g. brick, tile, cob, adobe, rammed earth). Matt black surfaces absorb the most heat but – unless you're a Goth – can be depressing to live in.

Solar heating can be 'active' – involving pumps or fans and electronic controls – or 'passive' – where the building does the work, with the minimum of operational input (e.g. you only need to open and close curtains or shutters, and vents or doors). Simple, intuitive operation always has less to go wrong – and no mechanical parts to wear out.

Solar heating works all year, so sun-rooms can get too hot. To prevent this, they need summer shade

possibilities (e.g. bamboo or Venetian blinds or, better, external awnings) and generous ventilation openings, especially high up. Indeed, a sun-room can even act as a solar chimney to cool the rest of the house if ventilation openings are placed well. On still days, air warmed in the sun-room will leave through high-level outlets, drawing cool air through the house from low-level inlets in its north side. (If it's windy, wind direction will determine how air moves, but the wind will anyway cool the house.) So, despite their bad reputation as energy drains that need heating in winter and cooling in summer,[17] conservatories can, in fact, offer both winter heating and summer cooling.

Mechanical systems	Passive systems
Electrically powered	The building does the work.
Rely on power-supply and equipment	Rely only on you.
Need maintenance and replacement	Little to go wrong.

Keypoints

- Glass walls admit winter sun. Glass roofs freeze us in winter, fry us in summer.
- Sunspaces / winter-gardens / conservatories should have no other heating or they'll waste more heat than they gain.
- Orientate these (or their glazed façades) within 30⁰ of south, if possible.
- Locate thermal mass where it's sun-warmed: it should be (moderately) dark-toned.
- Solar-heating systems should be 'passive', simple and intuitive to operate.

Choices

Options	Decision-making factors		
Sun-room for solar heat, light or semi-indoor greenery?	Primarily for:	Heat	Orientation is critical.
		Light	Orientation is less critical, as translucent-insulation roofing and night time insulation (e.g. shutters, blinds) can reduce heat loss.
		Greenery	Optimum orientation is desirable, but not critical.
South, south-west or 'nook' sun-room?	Length of heating season; period when heat needed: midwinter (due south), late spring (south-west is sometimes warmer); wind cooling; reflection from flanking walls / snow / water.		
South-west or west shade?	Length/severity of overheating season.		
Fixed, adjustable or foliage shade?	Predictability of summer weather.		
What ventilation against overheating?	Predictability of summer wind direction; height of potential 'solar chimney'.		

Resources

Information

Cotterell, J. and Dadeby, A. (2012) *PassivHaus Handbook*, Green Books, Cambridge.

Griffiths, N. (2012) *Eco-house Manual*, Haynes Publishing, Yeovil.

Hall, K. (ed.) (2008) *Green Building Bible*, Vol. 2, Green Building Press, Llandysul.

Nightingale, S. (2013) *The Woodstove Handbook*, Posthouse Publishing, Norfolk.

www.cse.org.uk/pdf/advice_leaflet_flue_gas_heat_recovery.pdf

Less common products

www.ceramicstove.com

www.discreteheat.com

1. Sammaljärvi, E. (1987), in *Det Sunda Huset*, Byggforsknin-grådet, Stockholm, page 120. Even if you regularly clean the filters, forced air is dry and contains few (if any) negative ions.
2. Flue gas heat recovery can save 37% of heating energy: *Green Building* magazine, Spring 2014, Vol. 23, No. 4.
3. *www.cse.org.uk/pdf/advice_leaflet_flue_gas_heat_recovery.pdf* (accessed: 7 July 2014).
4. Although in some countries most electricity is hydro-generated, 67% of the world's is fossil-fuel generated: Triodos Bank (2013) *The Colour of Money*, Autumn 2013
5. Griffiths, N. (2012) *Eco-house Manual*, Haynes Publishing, Yeovil; Cotterell, J. and Dadeby, A. (2012) *PassivHaus Handbook*, Green Books, Cambridge.
6. Some stoves allegedly don't need one. Personally, I wouldn't take that risk.
7. Some stoves have a second air inlet for this purpose.
8. Roaf, S. (2013) Flue fires in high performance housing: be prepared, *Green Building* magazine, Summer 2013, Vol. 23, No.1.
9. Roaf, S. (2010) Crossing the biomass chasm, *Green Building* magazine, Winter 2010, Vol. 20, No. 3.
10. von Zschock, R. (1996) Ceramic stoves, *Permaculture Magazine*, No. 13. Despite its (archetypally attractive) scent, wood-smoke is unhealthy. Advanced filters can reduce smoke-particulate by over 95%: Gates, D. (2010) Biomass in Scotland, *Green Building* magazine, Winter 2010, Vol. 20, No. 3.
11. Woodchip stoves are usually commercial-building scale. But many wood pellet ones are domestic-scaled.
12. Comfortable room temperatures are normally considered to be in the 20-24°C range (68-75°F), although clothing, activity, health and humidity, air movement and radiation affect this.
13. Check their wattage: much of this power ends up as heat.
14. From my experience with boiler breakdowns in icy weather, 60cm (2') internal stone walls keep temperatures just tolerable for a week. They take almost as long, however, to warm up again.
15. Solid to liquid phase-change at little above room temperature.
16. Hall, K. (ed.) (2008) *Green Building Bible*, Vol. 2, Green Building Press, Llandysul.
17. Around 90% are heated in winter; and many are also chilled in summer: Desai, P. (2010) *One Planet Communities*, John Wiley & Sons, Hoboken, NJ.

Keeping dry

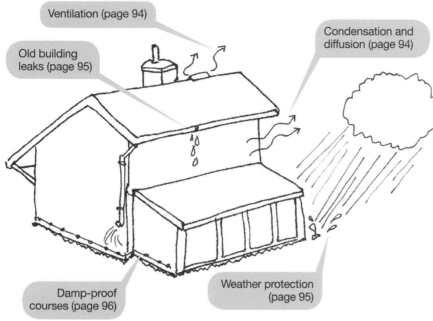

Ventilation (page 94)

Condensation and
diffusion (page 94)

Old building
leaks (page 95)

Damp-proof
courses (page 96)

Weather protection
(page 95)

Dampness and dryness basics

As moist air conducts heat better than dry air, in damp climates, air temperatures tolerable elsewhere chill us to the bone. Indeed, in moderate climates (or seasons), houses often only need heating because the air is damp. Were it dry, they'd feel warm enough.

All air contains some water vapour, but indoors we add lots by breathing, cooking, showering and especially clothes drying. Humidity, in fact, is the most common indoor pollutant in most houses. As water vapour is harmless to breathe and steam helps clear up bronchial coughs, it may not sound polluting, but its effects are unhealthy. Dust mites can't live below 50% RH, but 70-83% RH and 25°C (77°F) temperature is heaven for them.[1] Fungi also depend on moisture – often from condensation. They can destroy buildings and their spores are unhealthy to breathe. Formaldehyde is water-soluble, so humidity increases its release (e.g. from glues, chipboard, MDF). Also, insulation is much less effective when damp (except for wool, which actually warms slightly when wet). Damp insulation increases heat loss, which in turn makes walls colder, increasing condensation, which makes the insulation wetter still… a vicious spiral.

Water content and relative humidity of air

The air we breathe isn't just oxygen, nitrogen, CO_2 and trace gases. It also contains water vapour. But there's a difference between water content – which can be extracted and measured – and relative humidity (RH) – which we experience as 'humidity'. Warm air can hold more water vapour than cold air, so as air cools, water condenses out. We see this as clouds, fog or condensation droplets on cold, impervious surfaces. The air's water content divided by the amount it could contain at its current temperature is its relative humidity. Irrespective of water content, as temperature falls, this RH figure rises; and as temperature rises, it falls. Cold winter air, therefore, holds little water; and dry-feeling warm summer – or indoor – air can hold a lot.

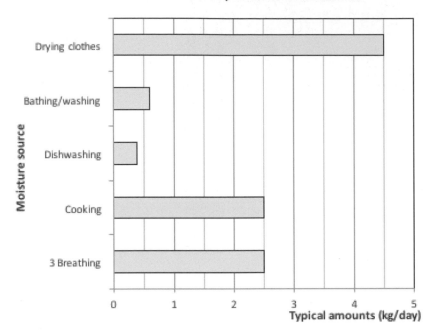

Moisture: the most common air pollutants in most homes[2]

For all this, the answer is fresh air. This dilutes and replaces damp room air and, especially if it's warm, draws moisture out of damp materials. Lots of fresh air, however, means lots of heat lost. It's better, therefore, to keep air dry by ensuring we don't *add* much moisture. In addition to minimising added moisture and ventilation to remove this, the materials you build with can help regulate humidity (see: **Design for dryness**, page 94).

Another source of moisture is rising damp. Modern buildings routinely have damp-proof courses (DPCs). Few pre-twentieth-century ones do, although some had courses of slates, tiles or engineering bricks, which work almost as well. As capillarity draws groundwater upwards, both through soil and into building fabric, rising damp is always a risk, and is common in old buildings. Different soils can hold different amounts of water, and subterranean water tables vary in depth. Gravel and sand are free-draining. Humus and clay absorb and retain moisture. (Clay, however, swells when wet, making an impermeable barrier. This can prevent up-flow of water below it or hold water above it. It's so effective that it's used to line canals and to core earth dams.)

How materials get wet, and stay wet or dry out

- Capillarity: solid materials filled with narrow spaces can take up liquids through 'capillary action': intermolecular forces between the liquid and the solid material enable the liquid to travel upwards. Also known as 'wicking'. This can cause rising damp.

- Hygroscopicity[†]: the ability of a material to attract and hold water molecules within it. Hygroscopic materials like clay are good at regulating indoor air humidity: they absorb water vapour from the air and either store or dissipate it, depending on conditions.

- Vapour-permeability: the extent to which air – moist or dry – can pass through a material. This can either increase or dissipate the moisture-content of the material, depending on conditions, so buffering the humidity of indoor air.

- As it's almost impossible to keep moisture out, it's vital it can *get* out.

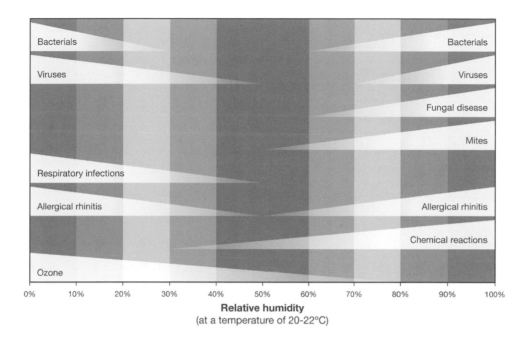

Relative humidity
(at a temperature of 20-22°C)

Additionally, lack of maintenance over the years means that many older buildings also suffer from leaks, gutter problems and rain-splash. Fortunately, most of such faults are easy to remedy (see: **Making an old building dry**, page 95).

The building itself is a source of dampness. New buildings, constructed with 'wet trades' (e.g. brickwork, concrete, plaster), take some two years to dry out (but see: **Chapter 7: Airtight construction**, page 70 for wet-plaster's benefits). Materials allowed to get wet when stored outdoors transmit this moisture to indoor air and also further dampen the fabric.

At the other extreme, air can be *too* dry. We generally feel most comfortable in air with 40-60% RH – which is also the optimum humidity for health. In cold winters, outdoor air has had the moisture freeze-dried out and then is further dried by warmth inside. Air-conditioning also dries air. Excessively dry air can cause sore throats. It also increases materials' resistance to electricity. This means we build up static charges (especially on synthetic carpets) and can release them with painful (although harmless) electric shocks. (Air-conditioned offices, combining refrigeration-dried air with synthetic furnishings upon which we rub, are particularly prone to this.) Again, wood and clay have

a humidity-moderating capacity. But it may also be necessary to add moisture (e.g. by evaporating water from a bowl on or beside a radiator or stove; or from well-watered indoor plants or a water feature).

Keypoints

- Damp can damage building fabric. Air that is too damp, or (less commonly) too dry, can cause health problems.

- Indoors, we add lots of water vapour by breathing, cooking, showering and, especially, clothes drying.

- Fresh air dilutes and replaces damp room air and, if warm, draws moisture out of damp materials.

- Lots of fresh air, however, means lots of heat lost.

- Hygroscopic materials moderate humidity.

- Few older buildings have DPCs. Many have rising damp. Many also suffer from leaks, defective design and/or construction and deteriorating fabric and/or rainwater goods.

- New wet-trade-constructed buildings take time to dry out.

- If dry air causes static-electric shocks, dry throats and/or skin rashes, it needs rehumidifying.

Design for dryness

Although dryness can be a problem, in moderate climates indoor humidity is a more common one. The most important action to prevent this is to minimise the amount of water you add to indoor air. Since drying laundry is the biggest contributor to indoor humidity, you may be tempted to turn to laundry driers – don't; these use *lots* of electricity. Instead, you need outdoor clothes-drying facilities. When it's raining, use a porch, draught-lobby, greenhouse or conservatory with windows open. (Despite lasting longer and costing nothing to use, these cost more to build than laundry driers cost to buy, so developers rarely provide them.)

Ventilation is the next most important design consideration for reducing dampness. Moisture-producing rooms (e.g. bathrooms, kitchens) need ventilation extracts. This needn't involve noisy (and power-consuming) fans but can use 'passive ventilation'[†]. To save needless warm air loss, outlets can be automatically (non-electrically) controlled to only open when indoor air is humid. Passive ventilation works best when ducts (which are normally 125mm: 5in diameter) have few bends and their terminals are high (at least 1.5m/5' above the ceiling) and in the *suction* zone above roofs. As wind direction varies, this usually means at (or near to, and above) the roof ridge. It's inadvisable to join up separate ducts due to cross-contamination risk. If built into chimneys, ducts are also warmed, which accelerates upward airflow.

Inadequate ventilation means humidity builds up, increasing risk of condensation on any cooler surface (e.g. within built-in cupboards, on window reveals and walls behind furniture or at north-east corners). Condensation is followed by mould – which is unsightly, odiferous and seriously *unhealthy*. In extreme cases, condensation can cause timber-destroying rot. However, even surface mould is no trivial matter: I've seen apartment walls black with mould, wholly unfit for habitation – but solely due to lack of ventilation.

Although adequate ventilation is the most effective way to control humidity, the materials you build with can also help regulate it. Hygroscopic materials absorb humidity, and then release it under dry condi-

For passive ventilation, minimise bends in ducts and place outlets near the roof ridge.

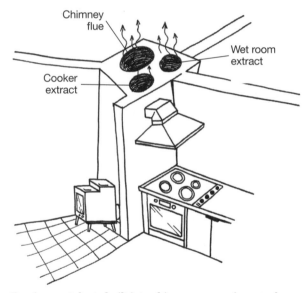

Passive vent ducts built into chimneys: warming accelerates airflow.

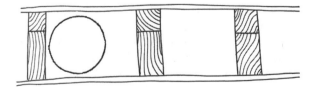

Passive vent ducts can be concealed in slightly widened upstairs partition walls (seen here in plan).

tions. The best of these is clay – which dries before our eyes. (Clay plaster suffices; unfired clay bricks are even better – besides causing minimal CO_2 emissions to manufacture.) Wood, lime and wool are also good and even masonry does a little (lime-mortared brickwork and stonework and, particularly, calcium silicate (sand-lime) bricks,[4] which also have substantial thermal mass. Concrete, however, dries too slowly.). Some of these materials also improve air quality in additional ways: wood, ex-plants and unfired clay remove toxins from the air, and lime is bactericidal. (Hence the use of spun-peat clothing against air pollution and radioactivity;[5] and charcoal, kaolin and chalk as gut purifiers.) To work, they must, of course, be *unsealed*: no oil-based or plastic paints (e.g. oil-paint, acrylic emulsion) or varnishes (e.g. polyurethane). This therefore excludes surfaces that must be wipe-cleanable for hygiene.

Hygroscopic building materials include (in order of effectiveness):
- clay;
- wood;
- plant and animal fibres (e.g. wool, silk, cotton, peat);
- lime;
- sand-lime bricks;
- even masonry if not sealed by acrylic paint.

Leaks, driving rain, punctured or imperfectly sealed vapour-barriers and frequent soaking (e.g. from rain-splash) can make walls and roofs damp. As it's almost impossible to *completely* prevent at least some water getting in like this, ensure the resulting dampness can get out. If not, there'll be rot and mould. It's safer, therefore, to make construction *slightly* permeable, so it can dry out (see: **Chapter 6: Building physics**, page 56). Similarly, masonry walls need vapour-diffusing materials (e.g. brick, not concrete, or unventilated plastic cladding) on the outside. Breathability not only helps walls dry, but also lets indoor pollution out and fresh (but filtered) air in (also, although it's *not* the same thing, see: **Chapter 7: Dynamic insulation**, page 68). Just as wearing breathable fabric feels more comfortable than wearing impervious plastic, buildings of breathable-construction are healthier and more pleasant to live in.

Keypoints

- Humid air narrows comfort temperature range; dry air broadens it.
- Add moisture to excessively cold-dried air.
- Inadequate ventilation increases risk of condensation, mould and rot.
- All moisture-producing rooms need ventilation extracts.
- 'Passive ventilation' ducts should have few bends and (preferably) ridge outlets.
- Assume (some) water will always get in: it's therefore essential it can get *out*.
- Breathability helps walls dry.

Choices

Options	Decision-making factors
More ventilation or warmer surfaces?	Energy conservation; reducible, remediable or unavoidable moisture source?
Adjustable, passive, trickle or extract-fan ventilation? In which rooms?	Moisture source; fan noise; can passive vent outlet(s) be high enough above the ceiling? Will you adjust manually operated ventilators?
Which rooms need hygroscopic-material surfaces? How hygroscopic?	Moisture source; wipe-down / washable surfaces?

Making an old building dry

Many old buildings are damp, but why? Is the cause easily remediable? (It's usually easier than structural faults.) Dampness could be due to a variety of causes including plumbing leaks; roof leaks; flashings and chimneys; gutter leaks or blockages; rainwater downpipes so short that they discharge on to the wall; driving rain; condensation; things built (e.g. garden walls), or temporally leant, against the house walls; cracks in masonry and rendering; rain-splash off paving; or rain running back under windowsills. These fall into three categories: easy repairs and/or remedies, difficult (or difficult to access) repairs, and rising damp.

Autumn leaves fill gutters every year. If there are enough, they can bridge back to the roof and/or block downpipes. Similarly, any ball that reaches a sloping roof will roll down into the gutter then float to the outlet. All of the rainwater that falls on that side of the roof can then wash, or be blown, on to the wall. Clearing gutters is a simple job (but see: **Chapter 19: You build your house**, page 223 for ladder safety). Over-short downpipes soak walls in the same way, but are easy to get to. All you need do is extend the pipe or, if you can't joint a bit on, sleeve its end in a larger pipe. Things leant against house walls may seem inconsequential, but off them rain splashes on to the wall, and behind them walls have no chance to dry out. Such causes of damp are obvious and easy to remedy, but surprisingly common.

Leaking roofs and chimney flashings are more difficult; and plumbing leaks are more difficult again, as they're often in inaccessible places and require fittings you don't have to hand. None of these (except complex lead dressing), however, require more than basic DIY skills.

One of the most common causes of dampness in old buildings, however, is rising damp. This is the natural process by which anything porous (i.e. almost all building materials except metals, plastics and glass) takes up water from the ground by capillary action. It's only found in masonry and earthen walls; any timber buildings built without separation from damp ground will already have rotted. Damp rarely rises above 4' (1.2m) within an exposed wall, before air dries it out.[6] (It can rise more if walls are thick, absorbent or have vapour-sealed faces; also, because plaster is porous, it'll rise further up the plaster faces than inside the wall itself.[7]) At its maximum-extent contour, all the soluble salts are deposited – which can destroy surface finishes, or sometimes even walls themselves. This can be a greater problem than the damp itself, which only needs lots of air to keep it mould-free.

Some pre-twentieth-century houses have severe rising damp but many don't. If there's only a little, it may not be serious enough to require installing a DPC. (The meters used by DPC-installers' surveyors can usually find damp – but, as they're actually salesmen and can exploit the seriousness of damp to scare you, you can't always trust their advice. Indeed, a *Which* investigation found that almost half DPC-installation recommendations were unnecessary, if not fraudulent.[8])

Unoccupied buildings are often closed up and unheated, so prone to condensation. Like all damp, this soaks into their fabric (masonry and wood) and, so long as it's unsealed, will dry out if well ventilated. (Sealing doesn't protect; it keeps moisture *in* – where it does most damage. You can see this clearly wherever polyurethane-varnished wood is exposed to rain for a while.)

If a house seems only mildly damp, but dry enough to live in, does the lack of a DPC matter? If you're planning to improve the insulation and airtightness, these will change the rate at which walls can dry, so it well may. If you'll add *external* insulation, you'll almost certainly need to install a DPC, as impermeable plastic foams keep all moisture in; and even breathable insulation will retard the walls' drying rate. Similarly, if you draught-proof thoroughly, interior surfaces won't get enough air to dry fast, which increases the likelihood of mould. For *internal* insulation, painting the interior face of the wall with a bitumen-based waterproofing solution before fitting this may suffice. (With anything chemical, you must be sure that materials don't attack each other; e.g. bitumen solvents destroy polythene.) This waterproofing solution will need to extend well above, and to the side of, the current rising damp patch, and along abutting interior walls, because damp that now can't get out will rise further and extend further. It may also be necessary to increase the ventilation rate (e.g. with humidistat-controlled passive vents). Whether the economy of not retrofitting a DPC is worth the risk, however, is arguable. Personally, I usually err on the side of caution.

For retrofitting damp-proof courses, there are five methods:

- Physical damp-proof courses inserted into slots sawn through masonry.
- Injected chemical damp-proof courses.
- Ceramic tubes (or special bricks) inserted to drain damp air (cold and heavy) out of masonry and replace it with (lighter) dry air, so drying the wall.
- Earthed copper strip to ionise moisture so it doesn't rise by capillarity (e.g. a former Rentokil method).
- A non-electrical device that counters capillarity by creating an electrostatic field over a whole area, which diverts water flow (e.g. the Aquapol method).

All of these have their advocates and detractors. Some methods are obviously more suitable to some circumstances than others.

The first one (a physical barrier) provides a tangible water stop, but is hard, slow, noisy and dusty work: the slot can't be sawn all at once or there's nothing to hold the building up. It best suits narrow brick walls. The last one, based on electrostatics, sounds like magic but works well enough to be used in many famous European historic buildings. Injected chemical DPCs are the most common but can be compromised by air voids and settlement in rubble-cored walls; and their solvent fumes pollute indoor air for few weeks. (As this method involves specialist equipment, it's usually undertaken by specialists, although the skills required are well within DIY range.) Copper strips and earthing rods appear to work but copper doesn't last forever in acid soil (perhaps 30 years). The ceramic tube system makes sense in theory but is widely discredited.[9]

Which is best? Physical DPC insertion obviously must work. I've lived with both copper strip and chemical injection systems; both worked tolerably well, although not perfectly (unventilated areas smelt of, but didn't show, mould). I trust the 'magic' one because of its prestigious client list. (It's also more expensive, so ought to be better.) But I don't know which is best; I'm not sure anyone does.

Keypoints

- Damp can come from rain, plumbing, condensation or groundwater.
- After mending leaks and installing horizontal barriers to rising damp or vertical barriers to damp abutments, air is essential to dry out building fabric.
- Traditional building materials absorb water then diffuse it. Sealing may slow absorption but – much more seriously – prevents drying out.
- Traditional DPCs come in several forms. So do remedial DPCs. So does the reliability of salesmen's advice.

Choices

Options	Decision-making factors
Install or don't install a retrofit DPC?	Mould risk; source of damp; severity; ventilation adequacy; timber rot risk; cost benefit; opportunity to install at a later date.
Which system?	Suitability to circumstance; disruption; solvent fumes; the system / installer's reputation; reliability of guarantees.

Resources

Information

BRE guide: Understanding Dampness.
Cotterell, J. and Dandeby, A. (2012) *PassivHaus Handbook*, Green Books, Cambridge.
Green Building Digest 20, Ventilation, Summer 1999, Queen's University, Belfast.
Oliver, A. (1988) *Dampness in Buildings*, Blackwell, Oxford
www.independent.co.uk/news/business/property-tubes-that-suck-up-your-money-but-not-the-damp-1163400.html
www.mybuilder.com/questions/v/953/do-you-need-to-remove-plaster-in-order-to-put-in-a-injected-damp-proof-course-alternatively-is-installing-solid-wall-insulation-at-the-same-time-a-good-idea

*www.which.co.uk/news/2011/12/damp-
proofing-companies-exposed-in-which-
investigation-274087*

Less common products and systems
Clay plaster: *www.clay-works.com*
Passive vents: *www.passivent.com*
Retrofit damp-proofing: *www.aquapol.co.uk*
Unfired clay bricks: *www.ibstock.com
sustainability-ecozone.asp*

....

1. Research at Wright State University, OH; Mason Hunter, L. (1990) *The Healthy Home,* Pocket Books, New York.
2. *Green Building Digest 20,* Ventilation, Summer 1999, Queen's University, Belfast.
3. Around 0.6 litres(1 pint)/person/day if inactive; much more if active.
4. Cotterell, J. and Dadeby, A. (2012) *PassivHaus Handbook,* Green Books, Cambridge.
5. *www.anthromed.org/Article.aspx?artpk=252* (accessed: 22 May 2014).
6. *www.mybuilder.com/questions/v/953/do-you-need-to-remove-plaster-in-order-to-put-in-a-injected-damp-proof-course-alternatively-is-installing-solid-wall-insulation-at-the-same-time-a-good-idea* (accessed: 12 December 2013).
7. *www.aquapol.co.uk/* (accessed: 12 December 2013).
8. *www.which.co.uk/news/2011/12/damp-proofing-companies-exposed-in-which-investigation-274087* (accessed: 17 May 2014)
9. *www.independent.co.uk/news/business/property-tubes-that-suck-up-your-money-but-not-the-damp-1163400.html* (accessed: 10 November 2013); BRE guide: Understanding dampness.

Keeping warm and dry: implications for different types of construction

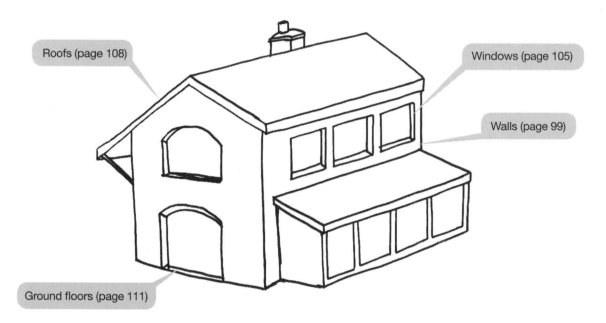

CHAPTER TEN

Roofs (page 108)

Windows (page 105)

Walls (page 99)

Ground floors (page 111)

Warm and dry walls

Walls have many functions. These include security, structure, insulation, thermal stability and, like clothes, projection of our (presented) personalities, all of which affect wall design, construction and material. However, we wouldn't build houses with walls unless they could keep us dry and (at least tolerably) warm. How walls keep us dry and warm varies with their construction. Consequently, so do the techniques for improving their insulation value.

Most older buildings are built of solid brick or stone, often with a rubble core. Most modern (post-1930) ones – some 80% of UK houses – are built of brick (and/or block) with cavity walls. Some are rendered (stuccoed); some, stone-faced. A very few houses – although many apartment blocks – are concrete-panelled. An increasing number of new houses *look like* brick, but this is just a facing on timber frames. Some houses are timber- or tile-clad on timber frames, some on brick. When it rains, concrete, brick, render, stone and even wood shed most of the water but absorb some – or, if it's wind-driven, a lot. When the rain stops, these dry out: mostly from their external faces, but also, if they're thin (e.g. the outer leaves of cavity walls, timber rain-screens) from interior faces. Much of London's housing, for instance, is built of 9" (225mm) brick. Most is acceptably dry – but unacceptably cold.

> Whether the interior faces of external walls are dry depends on four factors (besides construction shortcomings; see: **Chapter 9. Keeping dry**, page 91):
> - the wall's absorption and, particularly, *diffusion* rates;
> - its thickness;
> - the amount, duration and frequency of rain;
> - exposure to wind-driven rain.

How, where and with what material we insulate mustn't diminish walls' capacity to dry out. This is doubly important as a warmer climate can hold more moisture, so we must anticipate more, more persistent and more violent rain.

Masonry construction

Solid masonry dries from both sides, but for dry homes, we want it to dry from the outside before the inside gets damp. Similarly, cavity walls' outer leaves dry from both sides, but faster from outside as there's more air movement. In parts of Europe where winters are cold and snowy, wide multi-cavity bricks are common. Their multiple holes limit the depth to which moisture penetrates and their width provides durability, insulation, structure and thermal mass; all in one single-leaf wall. In Britain, where rain is frequent, cavity walls are common. In forested countries, timber con-

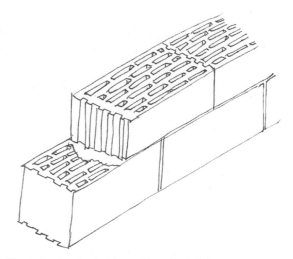

Single-leaf wall of wide multi-cavity bricks.

struction is traditional. Whereas solid walls are mono-layer (although serving multiple functions) and older cavity walls dual-layer, timber – and most modern – construction is *multi*-layer. In multi-layer walls, just as in birds' nests, the different layers have different functions. In birds' nests, these are structure (twigs), weatherproofing (grass, moss) and warmth (down).

What does a multi-layer approach in walls mean for construction?

In brick buildings, the outer leaf protects from rain and damage. The air-cavity prevents damp transmission. The insulation insulates. The inner leaf provides thermal mass, protects from damage and supports upper floors. Either or both leaves hold up the roof.

Different layers of construction: different functions

Where/what	Function
Outer leaf	Rain-screen (and sometimes roof support).
Cavity	Moisture barrier.
Insulation	Insulation.
Inner leaf	Thermal mass and floor (and sometimes roof) support.

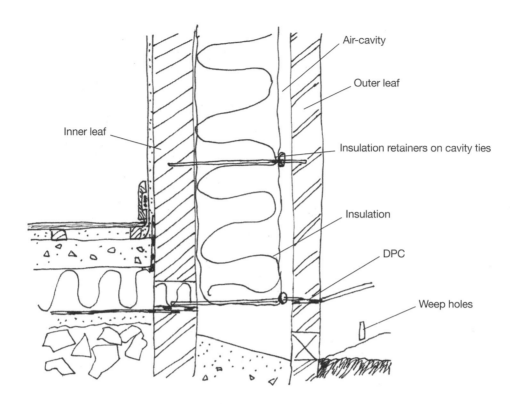

Optimum masonry cavity construction.

Timber construction

Many American and some European houses are built of wood, with timber, stucco or tile-hung cladding. (I'm not keen on stucco: it and wood move differently, so cracks are possible; and it *looks* like masonry but *is* fake; see: **Chpater 14: Soul/spirit factors**, page 173.) Wood-frame construction is also layered but the number of layers, and their materials and construction differ from those of masonry buildings. The outer layer (e.g. constructed of clapboard, vertical boarding with batten-covered expansion gaps, tile-hanging, etc.) sheds water. An airspace behind it, vented from base to roof-eaves[†], allows a drying airflow. Optimally, there should then be a vapour-diffusive wind-barrier (breather membrane[†] or moisture-resistant soft-board[†]), then the insulation between the structural uprights ('studs'). To stop the wall lozenging into a diagonal parallelogram, the structure is sheeted, usually with oriented-strand-board (OSB). Conventionally, this is on the outside of the studs, but on their inner face it also functions as a vapour-check. Then there should be a narrow cavity for cables so they don't puncture the vapour-check layer (OSB or membrane). To stop heat transmission through the studs, there should then be a further insulation layer, but this needs to be thin as it's on the room side of the vapour-check layer. For the interior wall surface (conventionally plasterboard), unfired clay bricks (with clay plaster) provide thermal mass and moderate humidity. Where there's a risk of earthquakes, wood-frame walls are the safest option (short of steel or heavily reinforced concrete),[1] as these can absorb seismic loads and, if the house does collapse, aren't heavy. Similarly, nails (which bend) are better than screws (which snap). And, of course, there should be no brick lining.

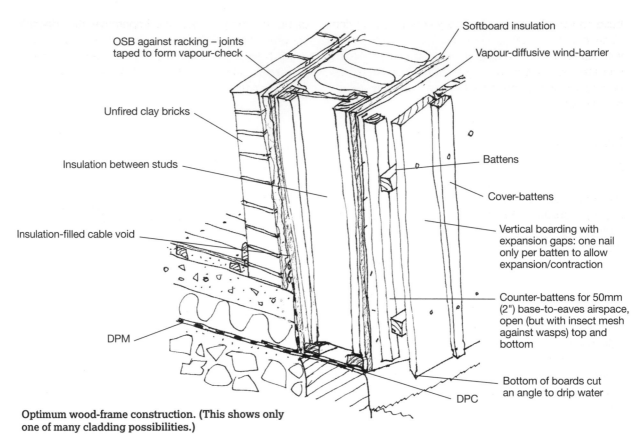

OSB against racking – joints taped to form vapour-check

Unfired clay bricks

Insulation between studs

Insulation-filled cable void

DPM

Softboard insulation

Vapour-diffusive wind-barrier

Battens

Cover-battens

Vertical boarding with expansion gaps: one nail only per batten to allow expansion/contraction

Counter-battens for 50mm (2") base-to-eaves airspace, open (but with insect mesh against wasps) top and bottom

Bottom of boards cut an angle to drip water

DPC

Optimum wood-frame construction. (This shows only one of many cladding possibilities.)

Earth or stone construction

Worldwide, the commonest forms of vernacular build-ings are of earth or stone. (Their external appearance is sometimes indistinguishable from rendered brick.) Although, in developed nations, earthen construction used to be considered inferior, it's enjoying a come-back. Of earthen techniques, cob is the most common – and easiest; rammed earth, the strongest and with the most thermal mass. Rammed earth, however, is made using shuttering (temporary wooden formwork), which is only economical if repeatedly reused – so uneconomic for single small buildings. Cob walls are normally 2' (600mm) wide. Although cob houses *feel* warm (so require less heating) because clay keeps indoor air dry, this doesn't provide enough insulation for modern regulations. Broader walls would. Cob is heavy, though, so broader walls entail lots more work – and longer drying-out time. One alternative is to

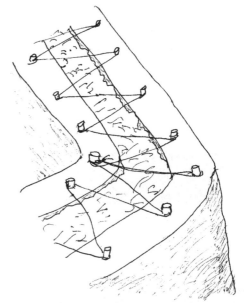

Cob-Leichtlehm-cob walls.

build cob or adobe faces and a lightweight, straw-rich straw-clay (Leichtlehm) core: the faces tied together with scraps of polypropylene baler-twine. Leichtlehm can also be used as a non-structural insulating infill between timber-frame members, or (as wattle-and-daub) over them.

Straw-bale construction

Another material currently experiencing a renaissance is straw, from straw-board (partitions) to straw-bale (external walls). Straw-bale walls provide good (and cheap) insulation. Structural straw-bale (Nebraska method) is draught-free, but settlement can open gaps around non-straw things, like window-frames. (Pre-compressing bales and post-compressing walls minimise, but can't be guaranteed to fully eliminate, this risk.) To ensure settlement doesn't compromise structure, straw bales are most commonly used as insulating infill between timber frameworks. There's risk, however, of differential movement of the timber and straw creating cracks, which admit draughts. To prevent this, the weatherproofing lime-based render must be separated from the wooden structure so it adheres only to the straw, not to the wood. Straw contains little nutrient, so doesn't easily rot but if it's

damp for long, it will. It's imperative this doesn't happen because, to get rid of mouldy straw, you would have to demolish the whole house.[2] I'm therefore cautious of using straw-bale construction in damp climates.[3]

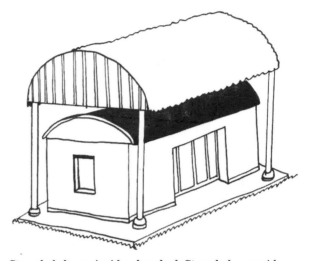

Straw-bale house inside a hayshed. Straw bales provide the insulation; the hayshed provides the structure. This separation of functions eliminates cold bridge risk and lets each material do what it does best.

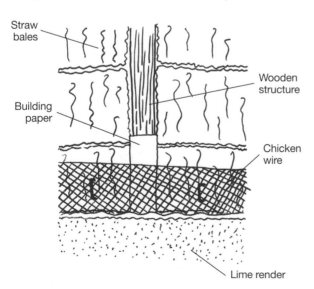

Straw bales

Building paper

Wooden structure

Chicken wire

Lime render

Although lime-render is somewhat flexible, as straw and wood move differently, there should be a separating layer (building paper) between it and the wooden structure.

Straw-bale house independent of roof structure. Separating structure and insulation eliminates cold bridges through the structural members. (It also limits settlement to that due to self-weight, so everything moves as one, reducing the risk of opening up draught-admitting gaps.)

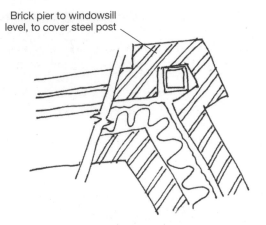

Brick pier to windowsill
level, to cover steel post

The same principle applied to conventional construction to eliminate cold bridges (plan).

Hempcrete

Hempcrete is a new material. Based on the same principle as Leichtlehm, it's a hemp-lime mixture: hemp-fibre for reinforcement and insulation, and lime for bonding. It can be cast in formwork or sprayed on to a screen-backed framework. It has excellent insulation and thermal mass characteristics, but isn't structural so needs a stud-frame. If you want to fix shelves and suchlike, this frame should be near the internal face. But if you'll be fixing a rain-screen to the outside, you need the frame (or fixing battens secured to it) near the external face. Like cob, however, hempcrete dries slowly: so work must be seasonally planned. A 1' (300mm) wall needs some 6-8 weeks before it can be lime-rendered. (If it's behind a rear-ventilated rain-screen, so dries faster, work can proceed faster.) As this material has only fairly recently been developed, I haven't used it, but (except for its long drying-out time and causticity-related hazards) hear nothing but praise for it.[4]

Issues for all construction types

There are too many other old, new, new-old or alternative techniques to list, but there are common issues, crucial to attend to. Wood, straw and all organic materials rot if wet for long;[5] and wet clay and earth soften and eventually disintegrate under load. The first concern, therefore, is to limit moisture entry: traditionally by giving the building a 'dry hat and boots': overhanging roof and stone base to limit contact with water. Traditionally, the base was either slate-topped or tall enough to dry out. In modern construction, it's DPC-topped and must be high enough to be above all rain-splash and plant growth. Above this, vulnerable materials are weather-protected. Even more important than preventing moisture ingress, is *preventing its retention*. Lime-render allows moisture to diffuse, and rear-ventilated rain-shields add airflow to further speed drying. Most rain-shields are wooden boards in one form or another, or, where driven rain is common, materials that shed water without getting wet (e.g. slates, tiles or even corrugated iron). Slates and tiles, however, must be securely fixed as wind can rattle loose ones off. They're also more vulnerable to impact damage than wood, so normally only clad upper floors.

All traditional construction is too soft to resist rats. They tunnel into lime mortar between stones and enjoy the warmth of clay and straw walls. Consequently, if you choose traditional soft construction, you need rat barriers. Traditional methods included a layer of broken glass atop the stone base wall, two courses of engineering brick (a traditional form of DPC) and stone 'mushrooms' (common under wooden granaries). Alternatively, you can use glaziers' offcuts or metal sheeting (as for termite shields, but not necessarily overhanging); or just cover the first 1'8" (500mm) of the wall in hard render, too smooth for claw purchase.

Whatever their material and construction, all external house-walls have a protective function. They protect against intruders, such as rats (and in the past, wolves, bears…) and, to different extents depending on circumstance, privacy invasion. These protective functions determine the position, type and extent of glazing. Additionally, as walls confront our line of vision, they influence our first impressions of a house: how welcoming or forbidding it is, and how respectful of, or unresponsive to, context. This affects choice of material, which in turn affects design.

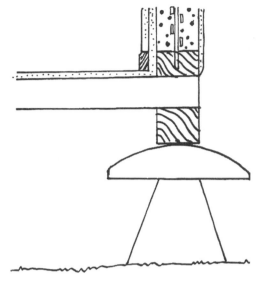

Traditional stone 'mushroom'.

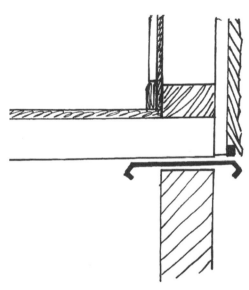

Termite shields made of glass or metal sheeting.

Keypoints

- Walls provide weather protection, insulation, thermal mass, structure, security and 'personality' projection.
- Their material and fenestration affects first impressions.
- Organic materials need 'dry hat and boots': protection from rain, splash and rising damp.
- Walls must be able to diffuse any moisture they absorb (which, however 'impervious' their materials, all except glass, metals and plastics will).
- Soft materials need protection from rats.

Choices

Options	Decision-making factors
Masonry, earthen, straw-bale, Hempcrete or wooden walls?	Local tradition; availability; available skills; humid or dry air; time (and weather) to dry out; climate for timber longevity; seismic risk.
Exposed brick, stone, render, timber boarding or tile- / slate-hanging?	Local tradition; availability; maintenance; impact damage; wind damage; driven rain; rot-encouraging or dry climate.
Wooden-board, slate or tile rain-shields?	Local tradition; impact damage; wind damage; driven rain; rot-encouraging or dry climate.

Warm windows

Windows and doors also need to provide the protective functions of walls but typically have much lower U-values than walls. External doors, and doors to cool rooms (e.g. larder and draught lobby) should be insulated when they're made – or you can add insulation and a matchboard lining. Windows lose heat – always at night and sometimes by day. New windows should, of course, be at least double-glazed with low-emissivity glass (also known as low-E glass, this reduces radiant heat loss or, reversed, solar-heat gain), insulating edge seals and a gas filling. (Argon is the cheapest

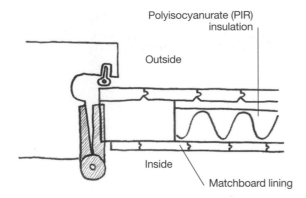

Insulating TG&V doors (tongue-and-groove with a V-joint): the 'V' where boards abut allows water to drain out.

insulating gas; xenon, the best-insulating. Krypton and xenon, however, are rare, so expensive and declining resources: I therefore don't recommend these.) Gases tend to slowly leak out, losing effectiveness.[6] Double-glazing panes should therefore be 16-18mm[7] apart so there's enough air to give adequate insulation once all the argon has gone. Triple-glazed windows give better insulation than double-glazed ones. They're currently 15-23% more expensive but prices are falling. In Central Europe, they're now only 7% more expensive.[8]

Even the best triple-glazing, however, has only about a tenth the insulation value of a well-insulated wall.[14] If this lowers temperature much more than 3°C (6°F) in one part of a room, it causes cold (thermally driven) draughts along the floor.[15] Moreover, in winter, most of the day is night, so radiation to dark sky, lower temperatures and no solar gain mean even more heat loss through windows – and all at a time when there's little to see outside but you don't want people looking in. The energy conservation implications led to a (fortunately brief) fashion for windowless buildings. Although life without windows is intolerable (see: **Chapter 14: Daylight and mood**, page 165) it does save energy. We don't have to do without windows, though. The answer is mobile insulation. External shutters are traditional in European houses and internal ones in Georgian, Regency and Victorian British houses. Insulated shutters are even better. (Small translucent panes in these can show your house is occupied.) Alternatively, use insulated blinds and/or heavy – and multi-layer – curtains. Pelmet boxes, tight to the ceiling or wall, stop warm air dropping outside this insulating layer. The better draught-sealed, or draught-obstructing, this meeting is, the less warm air can escape past them, and less cool air can come in. Similarly, the better this insulating layer meets or laps the house insulation, the less heat leakage there'll be.

Windows: thermal performance and light transmission

Type	U-value[9]	Light transmission[10]		g-value[†] (solar heat admission)[11]	
		(normal glass)	(white glass[†])		
Single glazing	4.8	1	1.11	0.87	
Double glazing	2.8	0.8	1	0.77 (air-filled)	0.70 (argon-filled)
Low-E double glazing	2				
Triple glazing	1.5	0.64	0.9	0.6 (white glass)	
				0.5 (normal glass)	

Heat loss from severely wind-exposed north windows is practically double that from sheltered south ones.[12]
As triple-glazing reduces solar gain, double-glazing is often a better choice for south-facing windows.[13]

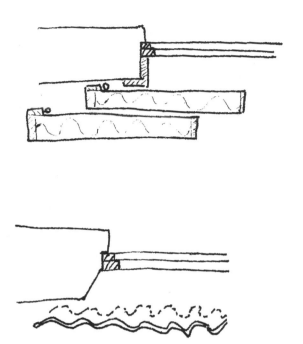

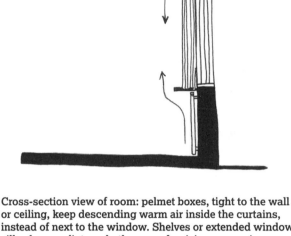

Mobile insulation: multi-layer curtains (bottom) and insulated internal shutters (top).

Cross-section view of room: pelmet boxes, tight to the wall or ceiling, keep descending warm air inside the curtains, instead of next to the window. Shelves or extended window-sills above radiators do the same for rising warm air.

Shutters and blinds can seal shut at night; and, on cold sunless days, be half closed. If you won't be home to close them, they can be motorised and automated. Automation, however, brings cat- (and baby -) trapping risk, so needs careful design (e.g. soft material, narrow inter-space to windows, pressure-sensitive cut-outs). Timer-based automation is generally too rigid for our flexible lives and variable weather. Light-sensitive automation can have a mind of its own: opening while you're dressing, and closing when you're doing something intricate. As you might guess, I don't trust machines to think about the full range of considerations. Unless there's a manual override, I don't trust them at all.

Internal shutters are easier to operate and immune from weathering deterioration, but require space to open, slide or fold into. (Shutters sliding behind shelving are accessible if they jam, whereas those sliding *into* walls aren't. Those folding on to walls or cupboard ends can be mirror-faced: more light in the room when open; and less radiant heat loss when

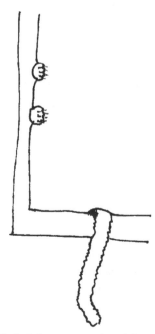

Automatic blinds bring cat-trapping risks.

closed.) External, draught-sealed insulated shutters make no demands on indoor space, but operating them can cause problems. You need either to open windows to shut/open them, or to use worm-gear that may jam or admit draughts, or electric controls which can go wrong (and a power-cut could black out both lamps *and* windows). Regardless of whether they're internal or external, draught-sealed insulated shutters can ensure windows lose no more heat than walls for the two-thirds of winter days that is night. This is a major energy saving.

Keypoints

- If possible, choose triple-glazed windows with low-E glass and argon filling.
- For solar heating, use white glass or double-glazing.
- Use mobile insulation (preferably insulated shutters) on all windows.
- Insulate and draught-seal external doors – and also doors to much cooler rooms.

Choices

Options	Decision-making factors
Double- or triple-glazing?	Climate; orientation; glazed area; insulation standard of rest of house; solar-heat admittance; cost.
Shutters, insulated blinds or heavy curtains?	Wall-space for shutters to slide or fold on to; views and light when part-closed; appearance; thermal efficiency; (thermally produced) draughts; things on windowsill.
Automated or manual operation?	Times when you're home; safety risks; convenience; controllability; reliability; electrical dependence.
Internal or external shutters?	Convenience; durability; space indoors; continuity with wall insulation layer; air-sealing; draught / cold air / operation issues; appearance; things on windowsill.

Insulating roofs: different constructions for different types

There are lots of kinds of roofs and lots of kinds of roof coverings, but only two broad categories: overlapping pieces of rain-shedding material (e.g. tiles, slates, shingles) and unbroken waterproof membranes (e.g. hot-bonded bituminous felt, glue-bonded sheets). Overlapping-bit roofs must slope – and slope at different angles depending on the materials/products used. Profiled metal sheeting can be laid at fairly shallow pitches (down to 1:48 for standing seam roofing[16]), whereas thatch must be very steep (e.g. 50-60°) as water remaining too long in the straw will rot it.[17] (It can't be too steep either, or straws slide out.) Slates must cover not only the gap between slates in the row below, but also their nail holes; and not only cover, but allow for wind-driven rain, both along and up the roof. Different sizes of slate, different degrees of exposure and use of stainless-steel clips instead of nails allow different minimum pitches. Tiles vary from shallow-pitch Roman and interlocking types to steep-pitch small plain tiles. Manufacturers list minimum pitches for their products but it's wise to always slope roofs a little steeper; and, if using a local material, never shallower than local practice. Past practice and standards suited past weather, but what you build now must cope with *future* weather.

Waterproof-membrane roofs can be any slope, down to almost flat (e.g. 1:40 slope[18]). Some waterproof membranes are made of PVC but EPDM synthetic rubber or rubber-based ones are more eco-friendly. To create living roofs, materials that are resistant to root secretions can be covered with planting medium (e.g. soil, expanded clay granules, seed-impregnated fibreboard). Most waterproof membranes need protecting from exploring roots by a root-inhibiting geo-textile membrane.[19] Unless you incorporate plastic netting for roots to grip or friction ribs, living roofs can't be too steep or rain-saturated soil will slide off. Around 25° is traditional but, with a scrap-carpet friction layer, spoilt hay to hold soil together and a spider-web of polypropylene baler-twine for roots to grip, I've made bits of roof at 45° – and they're still there![20] Since

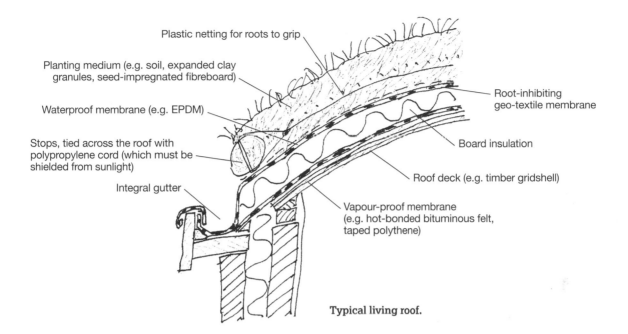

Plastic netting for roots to grip

Planting medium (e.g. soil, expanded clay granules, seed-impregnated fibreboard)

Waterproof membrane (e.g. EPDM)

Stops, tied across the roof with polypropylene cord (which must be shielded from sunlight)

Integral gutter

Root-inhibiting geo-textile membrane

Board insulation

Roof deck (e.g. timber gridshell)

Vapour-proof membrane (e.g. hot-bonded bituminous felt, taped polythene)

Typical living roof.

those days, proprietary living-roof systems have appeared. It's much safer to use one of these than improvise – as before 1990, I had to. Living roofs grow – so if sections of roof abut walls, the flashings must rise to a level above all possible vegetation. On poor, shallow soil (or expanded clay granule rooting layer), it's possible to limit vegetation to low-growing species (e.g. sedum). Deeper or richer soil grows taller plants, which will need weeding out. Since weeding often doesn't get done, it's safer to fit taller (hence more expensive) flashings instead.

The weatherproof layer affects a roof's slope and shape, but how it's insulated affects its design and construction. Insulated roofs are classified as warm or cold roofs. Cold roofs are those above the insulation layer. Pitched roofs above insulated flat ceilings are the most common examples. As condensation regularly occurs within the insulation, the unheated attic space must be ventilated to dry this out.

Warm roofs are those with the insulation above them. To prevent condensation within the insulation, there must be a vapour-proof membrane below it. As this must be guaranteed puncture-free, it usually doubles as the waterproofing layer. To prevent cold rain or snow-melt trickling through gaps between insulation

boards, there's usually a waterproof membrane above this. This needs either a semi-reflective face, so it doesn't melt or creep with heat, or a covering (e.g. a living green or brown roof[†]). With all this synthetic material, you might as well use a plastic insulation board (e.g. expanded polystyrene, PIR or PUR) but cork is a non-synthetic option. Warm roof construction suits flat, gridshell or concrete and ferro-cement roofs.

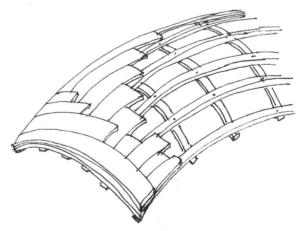

Gridshell roof: double-curved timber grid with two or three layers of boards screwed to it and/or each other.

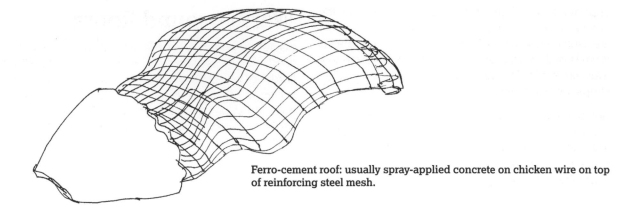

Ferro-cement roof: usually spray-applied concrete on chicken wire on top of reinforcing steel mesh.

Many roofs that are classified as 'warm' are actually 'semi-warm'. The insulation is between the rafters, often with softboard over the rafters, and/or an insulated ceiling below. Ordinary 100mm (4") rafters don't provide enough insulation depth, but engineered-timber I-beams can be of whatever depth you need. The depth required for insulation usually exceeds that required for structure, so deep rafters can span from eaves to ridge, allowing you to use a ridge-beam instead of a collared roof structure. This maximises space in the roof, increasing its potential usability.

Such semi-warm roofs need 50mm (2") unobstructed eave-to-eave ventilation above the insulation (and optimally a vented ridge), and a vapour-check membrane below it. This is usually polythene, but an intel-ligent membrane or bitumen-impregnated paper is preferable because, if punctured (as such membranes often are), its slight vapour permeability helps the insulation to dry. The five-to-one vapour permeability rule also applies here, so, instead of a membrane, you can use hardboard or OSB sheathing. As roofs get hot in summer, this accelerates formaldehyde off-gassing from OSB, so it's good to have the (otherwise unnecessary) vapour-check membrane below it, to prevent the formaldehyde gas entering the house. To prevent summer solar heat radiating into the room, you can use 5mm multifoil (or thicker, e.g. 12-40mm, if you want more insulation). As with walls, you also need a cable void to ensure nothing needs to penetrate the vapour-check membrane. This gives an

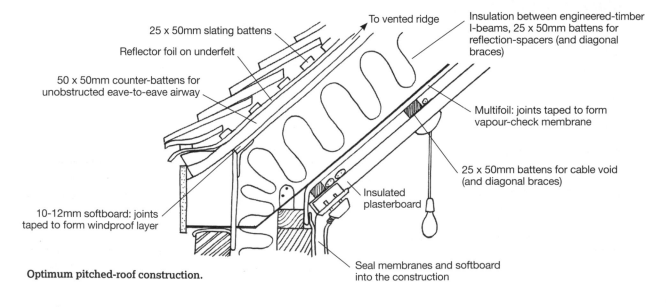

25 x 50mm slating battens

Reflector foil on underfelt

50 x 50mm counter-battens for unobstructed eave-to-eave airway

To vented ridge

Insulation between engineered-timber I-beams, 25 x 50mm battens for reflection-spacers (and diagonal braces)

Multifoil: joints taped to form vapour-check membrane

25 x 50mm battens for cable void (and diagonal braces)

Insulated plasterboard

10-12mm softboard: joints taped to form windproof layer

Seal membranes and softboard into the construction

Optimum pitched-roof construction.

opportunity to obstruct cold bridging through the beams with an insulation-lined ceiling. (Cable voids are usually formed by 25 x 50mm battens but, if headroom is limited, you can instead use 18mm battens or even 6mm plastic bottle tops; with galvanised steel straps for diagonal braces.)

Roofs need a vapour-permeable windproof membrane between insulation and airway. Both windproof and vapour-check membranes should be sealed (e.g. plastered) into the construction to be fully effective (see: **Chapter 7: Airtight construction**, page 70).

Keypoints

- Cold roofs need a ventilated airspace above the insulation.
- Warm roofs need a *completely* vapour-*proof* membrane below the insulation and (usually) a waterproof membrane above it. If the lower membrane isn't 100% vapour-proof, the moisture in indoor air will condense, compromising the insulation and forming bubbles in – and thus stressing, perhaps rupturing – the waterproof layer.
- Semi-warm roofs (i.e. insulated between rafters) need a vapour-check membrane below the insulation and a windproof vapour-permeable membrane and 50mm (2") ventilated airspace above it.
- Living roofs need a root-barrier to protect the waterproof membrane.

Choices

Options	Decision-making factors
Warm, cold or semi-warm roof?	Roof shape and slope(s); construction and material; space above; space for insulation.
Green (living), brown or conventional pitched roof?	Visual context; in or out of view; shape and slope(s); embodied pollution in membrane(s); weight; irrigation need/possibilities; stormwater attenuation need; abutment flashings.
Roof room(s) or unheated attic?	Headroom; need; stair/ladder position, headroom and convenience.

Warm ground floors

Insulating ground floors in new buildings is straightforward. The conventional way is to cover the underfloor hardcore with sand (so as not to puncture the DPM[†]), lay the DPM, lay expanded polystyrene on this, pour a concrete slab, then make the floor finish. Alternatively, fill the underfloor with expanded clay (e.g. Leca) or recycled-glass beads (e.g. Glapor), cap with lime-crete.[21] With this system, don't use a DPM as this would negate its breathability, and air between the beads denies water any capillary paths.[22] As always, however, you *must* separate any potentially damp masonry from timber with DPC or air gap.

Floor finishes on solid slabs are usually tiles, linoleum, or pressed bamboo sheet on a screed, or wood. (Don't use vinyl or laminates. Being electrical insulators, these build up electrostatic charges that attract dust, so the floor will never seem clean. And, worse, they off-gas.) Tiled floors are easy to maintain and won't be damaged by spilt water (or floodwater) but, unless carpeted, tiles are hard, so can tire or hurt your spine if you walk or stand on them for long. For a wooden floor, fix 50 x 50mm (2 x 2") battens to the solid base and fill between them with weak concrete or (better) lime mortar or limecrete – so there's no dead airspace (which encourages dry rot) – and fix floorboarding to the battens. This gives an opportunity to install underfloor heating/cooling – below the slab for constant, even temperature; or below the inter-batten mortar/concrete for quicker response to temperature controls.

Less conventional, but well-proven, newly revived floor finishes include end-grain wood (traditionally used under heavy machinery in Scandinavia) and linseed oil-sealed clay (traditional in the American Southwest). In dry countries, clay is durable; but in wet climates, although houses are dry indoors, too much water gets walked in from outside for this to be practical. End-grain floors can use timber offcuts (or de-barked logs), sanded down to make a level or – if you're adventurous – undulating floor, and then grouted.

The other, and probably most common, form of ground floor is a 'suspended' floor: floorboards laid on joists over a ventilated airspace. This construction particularly suits sloping sites, as you don't need to pack in hardcore fill to support a concrete slab. (If at all thick and inadequately compacted, fill can settle, causing the concrete to crack. Compaction, however, stresses the walls that retain the fill.) Suspended floors bring no such problems. Also, they're slightly resilient and can breathe, so are, in many ways, the best kind of floor to live with. To insulate suspended floors, staple bean netting to the underside of the joists, then lay in insulation batts[†]. Alternatively, nail battens to the joist sides and lay in softboard to make pockets for loose insulation (e.g. cellulose fibre), either pumped or hand-laid. Or form pockets by stapling breather (or kraft) paper to the joists. (Don't let the pocket sides come too far up the joists or every crinkle in the paper will let in draughts.)

If there's risk of rats chewing into any of these forms of insulation, you'll need to fix $^9/_{16}$" (13mm) chicken wire ('mouse wire') under the insulation (rats *can* chew through this to eat chickens, but probably won't bother just for a warm nest). For all suspended floors, ensure that fresh air can circulate to all parts of the underfloor: this requires at least 125mm (5") airspace beneath the joists and airbricks or grills every 1.8m (6') for through-flow of air. To keep this air dry, the underfloor airspace usually has a 50mm (2") concrete slab on polythene floor. And, as it mustn't fill with water, its lowermost part must be above ground, so it can drain. On level sites without enough space for ramps, this, however, can raise ground floors too high for disabled access.

Solid floors give more thermal mass, although this isn't optimally positioned. Suspended floors are cheaper to insulate, softer to walk on – and their wood anchors carbon. Both kinds, however, are easy to make warm. The only real complication with any kind of floor is ensuring that the underfloor insulation meets the wall insulation, so there's no cold bridge (see: **Warm and dry walls,** page 99).

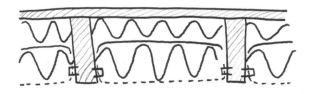

Netting stapled to the underside of the joists to support insulation batts.

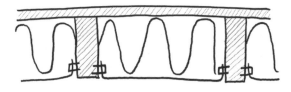

Breather (or kraft) paper stapled to joist sides as pockets for cellulose fibre.

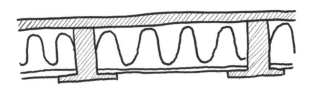

Battens nailed to joist sides to support softboard, then filled with insulation.

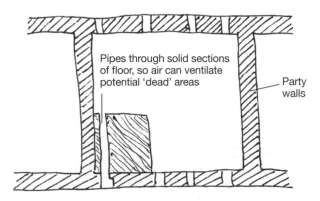

Pipes through solid sections of floor, so air can ventilate potential 'dead' areas

Party walls

Plan view of space under the ground floor. Airbricks and grills should be no more than 1.8m (6') apart: ensure these are unobstructed.

Keypoints

- Suspended floors can be insulated between the joists.
- They need an airspace beneath them, which must be through-vented and dry. If there's no space for a ramp, its depth may deny a house disabled access.
- Solid floors usually comprise a finish on a concrete slab, on top of insulation, on top of a DPM; or limecrete, granular insulation and no DPM.
- With Leca or glass-bead insulation, a DPM isn't normally necessary; with limecrete, it's undesirable.
- Underfloor heating/cooling piping can be below the slab, for stable temperature; or above it, covered in lime mortar / concrete, giving a quicker response.
- Solid floor finishes include tile, linoleum, bamboo and floorboards (with solid filling between their fixing battens).
- It's crucial that underfloor insulation and wall insulation meet to avoid cold bridges.

Choices

Options	Decision-making factors
Suspended or solid floor?	Disabled access; underfloor ventilation; need for thermal mass; flood risk; sloping or level site; washability .
Tile, linoleum, bamboo or wood floor?	Washability; hardness on spine; spillage risk; flood risk; climate (need for warm or cool floor); location/use (muddy footsteps); impact/wear resistance.
Deep or shallow underfloor heating piping?	Thermal stability or quick response to changing conditions?

Resources

Information

Cotterell, J. and Dadeby, A. (2012) *Passivhaus Handbook*, Green Books, Cambridge.
Evans, I., Smiley, L. and Smith, M. G. (2002) *The Hand-sculpted House: A Practical and Philosophical Guide to Building a Cob Cottage*, Chelsea Green Publishing, White River Junction, VT.
Guillaud, H. (1995) Earth construction technology: materials, techniques & know-how for new architectural achievements, CRATERRE.
Jones, B. (3rd ed. 2015) *Building with Straw Bales: A practical manual for self-builders and architects*, Green Books, Cambridge.
Minke, G. (2000) *Earth Construction Handbook*, WIT Press, Southampton.
Passive House+ journal, Temple Media, Dublin.
Steen, A. Swentzell, Steen B. and Bainbridge D. (1994) *The Straw Bale House*, Chelsea Green Publishing, White River Junction, VT.
www.bre.co.uk/filelibrary/pdf/projects/low_impact_materials/IP16_11.pdf

Less common products
For heat-conservation:
www.ancon.co.uk/products/wall-ties-and-restraint-fixings/cavity-wall-ties/ties-for-brick-to-block-construction
www.foamglas.co.uk
www.foamglas.co.uk/building/foamglas_perinsul
www.leca.ae
www.lime.org.uk/products/limecrete-and-sublime
www.limetechnology.co.uk/hemcrete.htm
www.marmox.co.uk/products/thermoblock

Living-roof systems and membranes
http://gbr.sarnafil.sika.com/en/group/sika-sarnafil/roofassured/singleplymembranes.html
www.alumascroofing.co.uk/products/waterproofing/bituminous-membranes/derbigum-flat-roof-membranes
www.enviromat.co.uk
www.greenroofsdirect.com/livingroofs.org
www.riefagreenroof.co.uk
www.rubberbond.co.uk
www.thermalcalconline.com
www.visionwaterproofing.co.uk/high-performance-epdm.htm

Wall materials
www.ceram.com/marss
www.lime.org.uk

1. Some argue that straw-bale, cob or reinforced masonry is safest. I'm not convinced.

2. In theory, straw doesn't rot below 84% RH, and only rots fast if at 98% RH for two days: Jones, B. (2nd ed. 2009) *Building with Straw Bales: A Practical Guide for the UK and Ireland,* Green Books, Totnes. Nonetheless, I'm still cautious.

3. Apparently, someone put a report on a straw-bale light-house on the internet as a joke. I have heard this cited as vindication of straw-bale in damp conditions – in an inter-national conference, no less! Lesson: don't trust what you read (even this book). Only trust what you've done or seen done – and only after sufficient time for faults to show up.

4. *www.limetechnology.co.uk/hemcrete.htm* (accessed: 9 June 2014).

5. Cork, coconut fibre and greenheart timber, however, are almost rot-proof.

6. Manufacturers claim 95% of argon will remain for 25 years: *www.double-glazing-info.com/Choosing-your-windows/Air-or-Argon-gap* (accessed: 13 July 2015). Others, however, consider they lose their effect after some 10 years: Cotterell, J. and Dadeby, A. (2012) *PassivHaus Handbook*, Green Books, Cambridge.

7. Cotterell and Dadeby recommend 20mm: ibid. But insulation peaks at 17mm and convection starts around 20mm: *www.greenbuildingforum.co.uk/newforum/comments.php?DiscussionID=125&Focus=10961* (accessed: 19 November 2014).

8. In Austria: Feist, W. (2013)Interview, *Passive House+*, 2013, Issue 5.

9. *www.designingbuildings.co.uk/wiki/U-value* (accessed: 29 November 2013).

10. Normal glass is 80% light-transmissive; 'white glass', 90%: Internorm Windows.

11. Cotterell and Dadeby, *PassivHaus*.

12. Fairweather, L. and Sliwa, J. (1977) *AJ Metric Handbook*, Architectural Press, London.

13. Hearne, J. (2014) Modern Galway home delivers ultra low energy bills, *Passive House+*, 2014, Issue 8.

14. Roaf, S. (2013) Resilient building design, *Green Building* magazine, Spring 2013, Vol. 22, No. 4.

15. Harvey, D. and Siddall, M. (2008) Advanced glazing systems and the economics of comfort, *Green Building* magazine, Spring 2008, Vol. 17, No. 4.

16. *www.metalconstruction.org/low_slope_roof* (accessed: 13 August 2015).

17. One thatching contractor, however, claims that 25° reed thatch can last up to 15 years (in a Dutch climate); but recommends 50°+ for a 45-year lifespan: *www.hiss-reet.com/constructions-with-reed/thatched-roof/thatched-roof-architecture/thatched-roof-design.html* (accessed: 3 September 2014).

18. 1:80 slope is normal, but I've seen too many leaking 'flat' roofs to trust this.

19. Some kinds are root resistant and need no further protection (e.g. see: *www.alumascroofing.co.uk/products/waterproofing/bituminous-membranes/derbigum-flat-roof-membranes*; *www.visionwaterproofing.co.uk/high-performance-epdm.htm* (accessed: 3 July 2014).

20. Twenty-five years ago, there weren't the materials available today.

21. Limecrete is carbon neutral, but cures slowly.

22. The Leca is usually left in its bags and the gaps between them filled with loose granules. Otherwise, when you come to pour the concrete/limecrete, you'll be wheel-ing wheelbarrows over a quaking bog – even if you use planks.

Keeping old buildings warm, cool and dry

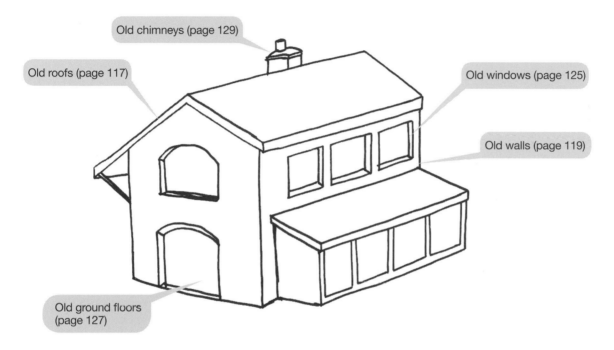

CHAPTER ELEVEN

Old chimneys (page 129)

Old roofs (page 117)

Old windows (page 125)

Old walls (page 119)

Old ground floors (page 127)

Insulating old buildings basics

To improve old buildings, insulation is fundamental. Unfortunately, however much formerly unsatisfactory old buildings are improved, they're still rarely as energy efficient as those designed for perfection from the outset. As British houses are among the most heat-leaky in Europe, they're definitely in the formerly unsatisfactory category and difficult to make as good as new eco-houses. You needn't feel bad about living in a less-than-eco old house, though. After all, if you have eco-improved it, it'll be much better than it was. Also, replacing energy-inefficient old buildings with energy-efficient new ones replaces operating energy spread over many years with an embodied-energy shock-load, so does little to reduce climate impacts (see: **Chapter 17: Materials we build with**, page 194). At needless great expense, it ignores the fact that replacing thermally-inefficient indoor clothing reduces climate impacts more than replacing thermally inefficient buildings (see: **Chapter 6: Heat loss basics**, page 47). Worse, it

Heat loss in a typical (UK) uninsulated house

Element	3-bed semi-detached house[1]	Bungalow[2]
Walls	35% (so 6% per wall per storey)	25% (6% per wall)
Roof	25%	30%
(Ground) floor	15%	7%
Windows	10%	13%
Draughts and air leakage	15%	20%

accords greater value to energy, which relates to convenience, than to cultural heritage, which relates to our humanness. Moreover, both the attitude behind demolition and the practice of covering up countryside with new buildings are inherently vandalistic.

Most old buildings lose massive amounts of heat. The first priority, therefore, is to identify where and how they lose it, and the easiest and most cost-effective interventions.

If less than perfection is inevitable, how much compromise is acceptable? Once you go to all the effort and expense of making semi-satisfactory improvements, you're unlikely to want to repeat the process to improve the result. Consequently, you get locked into 'second best'. To avoid 'lock-in', therefore, all energy conservation experts recommend 'do it once and do it right'. Easily said – but what if you can't afford to 'do it right'? As doing nothing means paying high energy bills (and causing lots of CO_2 production), this isn't satisfactory either. The best way round this is to strategise improvements: undertake them incrementally and either do each step 'right' or do it in such a way that it's easy to improve when you have more money.

What does avoiding 'lock-in' mean for incremental and partial improvements? With ground floors, for instance, second-hand carpet makes for warmer feet but doesn't lock you in to a less-than-satisfactory situation. Similarly, you can stop window draughts with sticky tape, removed in spring and replaced each winter. For windows themselves, improve the glazing on the largest, most wind-exposed or north-facing ones first. On the others, hang charity-shop blankets as curtains, make lift-off shutters, and/or improvise seasonal 'double-glazing' with cling-film (saran wrap) or bubble-wrap. What you *mustn't* do is spend lots of money on second best – which you'll only regret when energy becomes so expensive you can't afford adequate heating.

Usually, the first thing to properly insulate is the roof. This is where most heat per surface area is lost (or gained) because the roof is thin, exposed to the wind and night sky (or summer sun) and, because indoor air is thermally stratified, indoor–outdoor temperature differential is greatest here. Insulating roofs is also (usually) easy to do: typically only a few hours' work. Insulating walls is also (relatively) easy in theory, but not always in practice. Although only a small amount of wall surface is made up of windows, these lose a disproportionate amount of heat. Double-glazing is likewise simple in theory but, being thicker and heavier than single-glazing, can bring complications. Related to windows and doors are draughts (see: **Chapter 7: Airtight construction**, page 70). These both waste heat and cause discomfort. Draught-*sealing* can be difficult: wood moves with humidity so old construction is rarely airtight, and old window-frames don't allow room for inset seals. However, significant draught *reduction* is easy. It's the easiest, cheapest, most cost-effective and – after roof insulation – the most energy-conservative measure. As less fresh air means more condensation risk, however, it mustn't be considered a *substitute* for insulation, but as *complementary* (see: **Chapter 8: Keeping dry**, page 91). Most complicated and disruptive of all is insulating floors: you need to walk on them while you work, and, although the insulation itself is straightforward, raised floor levels mean lower ceilings and doorways. This makes a benefit/difficulty range, from the 'easier gains' of roof insulation to the disruptive complications of floor insulation. I therefore discuss these tasks in this order.

Insulating old roofs

Insulating conventional roofs is easy if you don't need the roof-space: kneeling on planks (so you don't fall through the ceiling), just unroll quilt between the joists, then over them, in the other direction. (If using mineral fibre, wear a dust mask, hat, overalls and gloves.) Ensure all gaps are filled. In Britain, aim for 300mm (12") total thickness (or, if you can insulate the rest of the house to PassivHaus standards, 500mm/20"). Leave an airspace around electrical cables[3] (or use over-capacity cable) and also around lamps recessed into the ceiling, so they don't get hot. (As you can't guarantee other people will use LED bulbs, this is an essential fire precaution.) Don't insulate under the cold-water tank, but wrap insulation *over* it and its pipes, so they don't freeze. Also insulate and draught-proof (e.g. tape) the ceiling hatch. It's vital, however, that you don't obstruct ventilation *above* the insulation, otherwise condensation water can't evaporate. (If there are gable vent grills, ventilation is already ensured. If these aren't, or can't be, installed, you'll need to fit eaves and ridge ventilators.)

If you're going to use the roof-space for storage, you'll need a deck over the insulation. To prevent crushing the insulation (which reduces its effectiveness), support the deck on stilts. Space-wise, you can probably accept the inconvenience of crouchingly low headroom and roof-trusses you must climb through. However, if you're making a room to live in, headroom is critical. This may restrict insulation depth, as the insulation must be above, not below, the room. As there must be a 50mm (2") unobstructed eave-to-eave airway above insulation (for which, you should ensure that the ridge-board and trimmers[†] around roof-lights and chimney don't obstruct airflow), this limits between-rafter insulation thickness. This may narrow your insulation options to the more efficient, hence slimmer but more expensive, materials. You may also need to restructure the roof (for which you'll need structural calculations for building regulations – not to mention safety); and/or slightly raise it (for which you'll need planning permission).

As roof rooms can overheat in summer, it's wise to incorporate a layer of heat-reflective foil – which you should position so that insulation can dry out (e.g. under tiling battens or as your vapour-check membrane). Under tiles, single-membrane reflective foil can drip condensation as, despite obstructing radiant heat transmission, it's a thermal conductor. It is therefore better to use thin (and cheapish) multifoil, which insulates (a little) against conductive heat loss.

If the roof room will be 'habitable', not just a storage attic, you'll need to be able to get out if there's a fire. (Indeed, there are regulations about this.) Several manufacturers make easy-to-climb-out-of roof-lights – but what after that? You need the roof-light to be where a ladder can reach it (so at the front if firefighters can't access your back garden), or where *you* can reach a flat (or safe-to-craw-on) roof, then an escape route. If the roof-space isn't a whole room, but just a sleeping loft, you can escape through the room it serves. Insulating roofs and converting the space within them for use isn't normally difficult: it just takes thought – and, sometimes more problematic, headroom.

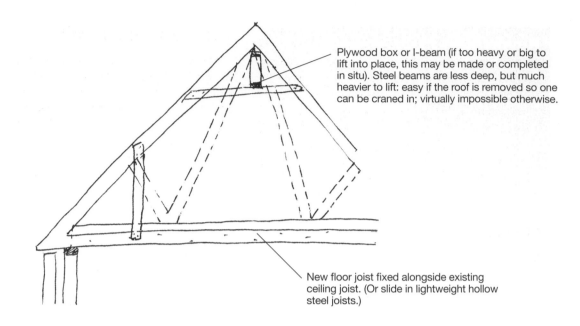

Plywood box or I-beam (if too heavy or big to lift into place, this may be made or completed in situ). Steel beams are less deep, but much heavier to lift: easy if the roof is removed so one can be craned in; virtually impossible otherwise.

New floor joist fixed alongside existing ceiling joist. (Or slide in lightweight hollow steel joists.)

Restructuring a trussed rafter roof.

Collar to ensure unobstructed airway at top

Quilt or PIR board between rafters, leaving an unobstructed 50mm (2") airway above; and filling purlin depth

Insulated plasterboard

Packings to thicken out to purlin depth

Hardboard strips as spacers

5mm multifoil (against summer sun), tape-sealed to form vapour-barrier

Existing purlin

Eaves ventilators

18mm battens and cable void.[4] Or you can fix plastic 6mm milkbottle-tops at 150mm (6") centres to the underside of the structure.

Minimum headroom roof insulation.

Choices

Options	Decision-making factors
Insulate at ceiling or roof level?	Space need; retention of future options; usability of space; feasibility of under- and/or between-rafter insulation.
Roof room or storage attic?	Space need; ladder/stair access; daylight possibilities; fire-escape possibilities.

Insulating old walls

Whereas new walls are easy and quick to insulate *during* construction, upgrading old ones isn't always straightforward. Most insulation retrofit is easy – but some is complicated; and some, almost inaccessible.

Timber-framed walls

For timber-framed buildings, existing walls can usually be insulated mostly within their structure, only widening them a little. With these, it's easy to avoid cold bridges. Thermal bypass, however, is hard to completely prevent, as access to install windproof membranes – and to seal these under battens – is often limited. Consequently, without stripping the wall down to its structural frame, the insulation won't be fully effective in windy weather.

Brick or block cavity walls

Brick or block cavity walls can be insulated by blowing insulation into the cavity. Such insulation, however, should always be a loose material (e.g. mineral wool, polystyrene beads) so it can be sucked out if any problems occur (e.g. caused by occasional high water-tables or driving rain – a problem more common in exposed coastal or mountain sites). Urea-formaldehyde foam emits formaldehyde, which is a health risk, so I avoid it. Cavity insulation makes a huge difference, but as cavities are normally only 2" (5cm),[5] this is far short of enough to make a house zero-energy. You may therefore want to add an additional insulation layer. If you can afford it, I recommend this. As the cavity insulation prevents cold bridges, this additional layer can be either internal or external.

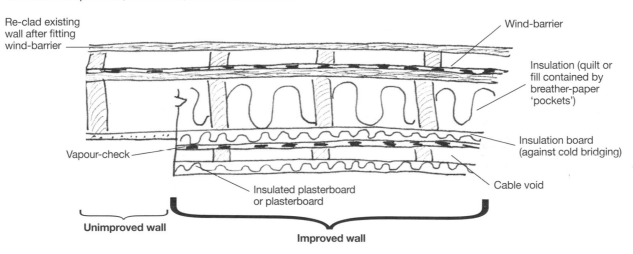

Re-clad existing wall after fitting wind-barrier

Wind-barrier

Insulation (quilt or fill contained by breather-paper 'pockets')

Insulation board (against cold bridging)

Vapour-check

Cable void

Insulated plasterboard or plasterboard

Unimproved wall

Improved wall

Timber-framed buildings can usually be insulated mostly within their structure, only widening walls a little.

Solid masonry walls

Solid masonry walls can be insulated internally or externally. These methods have different advantages and disadvantages; so suit different building uses, living patterns, climates, building construction and spatial and visual contexts.

Internal insulation

Internal insulation doesn't affect external appearance, so suits heritage buildings. It's protected from weather but stops buildings storing heat (or coolth), so lets interiors warm up fast – too fast in heatwaves. As both our warming climate and (increasingly likely) energy supply disruptions make thermal stability increasingly important, rapid thermal response is a serious disadvantage (unless homes are only intermittently occupied). Also, internal insulation shrinks rooms at least 50-75mm (2-3"). Thicker insulation would be better but rarely can the space be spared. (For barn and chapel conversions, where walls tend to be both cold and damp, there's often enough space for 100-300mm (4-12") mineral wool between these and a new block inner leaf: insulation *and* thermal mass.) If there's no space to spare but internal insulation is the only option (e.g. because the façade is attractive or you're bound by conservation zone restrictions), you can use vacuum panels. These, however, are very expensive (their price justified by a K-value of 0.004 watts/m.°K) and prevent fixing things to walls. Low-emissivity paint, although much less effective, is much cheaper. Being (like multifoil) a new product transferred from outer space, however, its efficacy in earthly weather is as yet unproven.[6]

Avoiding cold bridges to solid walls

The other main drawback of internal insulation is the potential for cold bridges. These can cause condensation, mould and even structural decay. Whereas cavity fill effectively eliminates this risk, with solid external walls, cold-bridge avoidance requires careful design. Wherever internally insulated solid external walls meet existing partition walls, cold bridges will occur. Either, therefore, cut the ends off (or remove and replace) the partitions, so the insulation can continue unbroken, or insulation-clad the partitions for *at least* 12" (300mm) from the junction to raise the cold bridge temperature above dewpoint.[7]

If you have enough space, the surest way to make an old building warm, dry and thermally stable is to build an insulation-wrapped home inside it. For this, chapels, barns and industrial buildings are usually wide enough; but houses and cowsheds rarely are.

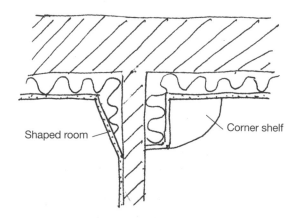
Shaped room   Corner shelf

To avoid cold bridges where interior walls meet external walls, return insulation for some way along them. This needn't produce an ungainly abrupt step, but can be visually 'absorbed' by angling the wall plane or fitting corner shelves.

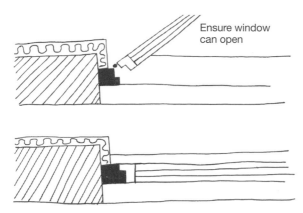

To prevent cold bridges, insulation (even if only thin) must return at doors and windows to lap their frames.

For both internal and external insulation, door and window reveals, soffits[†] and sills are potential cold bridges. The insulation should therefore clad such returns to overlap with door- and window-frames. To allow windows and doors to open fully, you may have to use thinner insulation (perhaps impermeable if it's higher performance) – but at least use *some*.

Avoiding cold-bridges where internally insulated walls meet floors

With floors, cold bridges are more serious. Although solid ground-floor insulation can easily join up with wall insulation (see: **Chapter 6: Avoiding common insulation failures**, page 62) what about where upper floors meet walls? Cold bridges here may not lose much *heat* but, as *condensation* within the floor depth isn't ventilated, it doesn't dry out. The ensuing damp can rot joists – or even spread dry rot throughout the building (including to neighbours' homes, so adding litigation to the already crippling remediation expense). You *must*, therefore, insulate wherever timber floors meet external walls. As there's risk of causing timbers to rot if you do it wrong, this requires close attention to detail. It's crucial there are no cold spots where condensation can form. There are three ways of internally insulating walls where they meet upper floors: improvising and hoping; ventilating potentially damp timber; and completely separating dry from damp. Doing it wrong doesn't *guarantee* you'll get dry rot, but makes it uncomfortably likely.

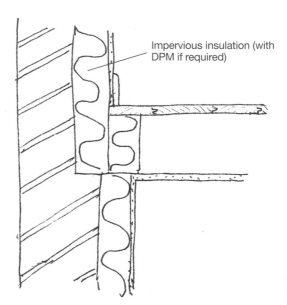

Avoiding cold bridges where floors meet walls: if joists run front-to-back, take off a floorboard to insert insulation.

If you use contractors, it's vital they understand and address this issue: some don't – or don't care (their job, after all, is to insulate, not protect from rot). If you do it yourself, this is rarely as simple as it might sound.

If joist ends are built into solid external walls, you can open up the ceiling and/or take off a floorboard, excavate the joist pocket to expose the joist ends, drill these and insert borate plugs, and also borate-paste then wrap them in DPC. Then tape-seal the wall insulation to the joists. This will keep the house warm, but leaves timber very near parts of the wall that could be damp (from condensation and/or rain). If the separation or insulation continuity isn't perfect, rot could start. As access is limited, this risk is high.

Alternatively, form a ventilated cavity *behind* the wall insulation[8] (having borate-plugged and -pasted joists, where they bridge this). This, of course, takes more room-space; and risks draughts if the timber shrinks. Although the timber may get damp, at least it has air to dry it, so probably suffers no more dampness than previously.

Safer than leaving joist ends close to damp walls, it's better to cut off and extract them, and seat the joists on hangers. To pull out ends, however, the pockets

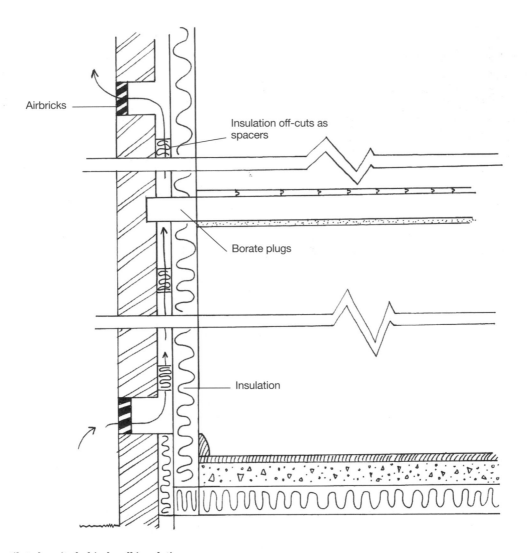

Airbricks

Insulation off-cuts as spacers

Borate plugs

Insulation

Ventilated cavity behind wall insulation.

they're in must be enlarged (or the remainder of each joist freed from floor and ceiling then moved out of the way). This turns a small job into a bigger one.

It's even safer to leave the joists exposed. There's both more air around them and if anything is going wrong, it's visible from the room below. Even with nail-scarred joists, this is often more attractive than flat ceilings. Such construction, however, will need to be fire-proofed (e.g. the joists intumescent painted[†],[9] and the underside of the floorboards plasterboarded). It also increases noise transmission between floors.

None of these methods, however, can *guarantee* timber won't rot. The only really safe way is to cut off and pull out all joist ends and support the (now short-ened) joists on a new beam. This doesn't have to be back-strainingly heavy steel. Alternatives include a lighter wooden beam, supported on posts (which you can use for shelf or cupboard sides), a timber-flanged metal-web beam, or a ply-webbed beam built in situ. If deeper than the floor depth, the beam needn't obstruct window-heads below. Instead, it can extend upward and be disguised as a ledge, seat or shelf. In whatever form, this method lets the insulation

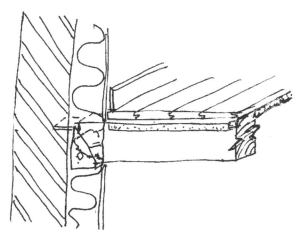

To keep joist ends away from damp walls and timber well ventilated: cut off joist ends; carry joists on hangers and borate-paste and DPC-wrap the (newly formed) ends buried in insulation. Fix plasterboard to underside of floor. (This, however, allows more inter-floor noise-transmission.)

continue without interruption. It *will* cost more, but there's nothing like doing something right to sleep well at night.

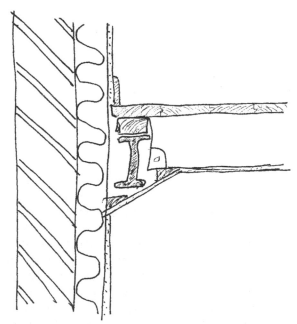

Best, but most expensive: cut off and pull out joist ends; carry joists on new steel beam. This is the only method guaranteed to be problem-free.

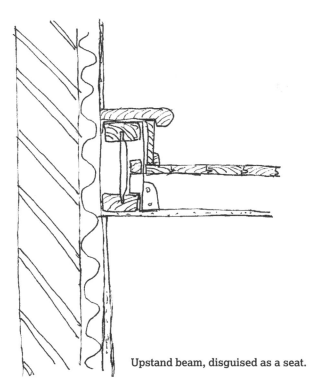

Upstand beam, disguised as a seat.

In terraced houses, if joists span between party walls, you can slip insulation in between the external wall and the joist beside it. If the gap is too narrow, you'll have to open up the pockets the joist ends sit in, prise off floorboards and cut out a strip of ceiling so you can slide the joist away from the wall. But will external damp or condensation continue into the party wall? This brings up the same problems as when joists run from front-to-back, although probably only for the outermost joist.

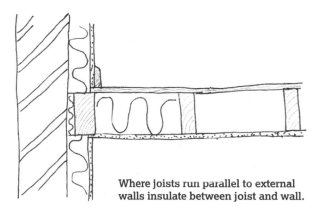

Where joists run parallel to external walls insulate between joist and wall.

External wall insulation

External insulation eliminates cold-bridges at upper-floor-to-wall junctions by keeping all existing walls within the insulated perimeter. This lets them act as thermal stores in both cold and hot weather: a significant benefit. Walls *must*, however, be dry; and the insulation must breathe, not be vapour-impermeable. Although simple in theory, installing external insulation can be complicated by external pipework and inadequate roof-overhangs. Additionally, as sucker or rooting tendril vines can destroy the insulation's thin protective render, any climbing plants must be twining tendril species, so need framework or wire support.

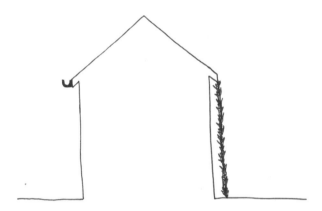

Over external insulation, climbing plants need frameworks/wires as sucker or rooting tendril species can destroy the insulation's protective render.

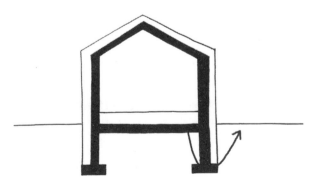

External insulation carried down to the foundations minimizes cold-bridging and partially compensates for inadequate floor insulation.

Because ground-floor floor-insulation is necessarily internal, external insulation leaves a cold bridge at ground level. Extending the wall insulation down to foundations can significantly compensate for both this and inadequate floor insulation. (When excavating, however, be mindful of underground services. Since electricity, gas, telecommunications, water and drains have connection points within the building, it should be possible to trace where these are.) Alternatively, on dry ground, you can extend the insulation outwards as a skirt, protected by paving. (Wet ground conducts heat too well for this to be worthwhile.) Despite such complications, if you're living in the house, external insulation is much less disruptive to install than internal insulation.

Insulation by building an extension

The easiest form of wall insulation for all types of building construction is to build an extension. To avoid cold bridges, insulation will often need to be overlapped or extended beyond the junction between extension and existing building. (For structural and constructional implications see: **Chapter 5: Structural problems: cracks**, page 41.)

For extensions, insulation may need to be extended to avoid cold bridges.

Choices

Options	Decision-making factors
External or internal insulation?	Cold bridge risks; room size; external appearance; disruption; thermal capacity; weather protection; impact and damage risks.
Build an extension?	Need; cost; space; location of poorest insulation / greatest heat loss.

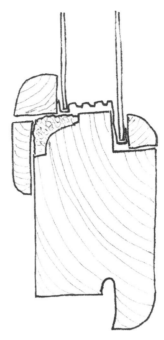

Double-glazing old single-glazed widows, using stepped units.

Wooden turnbuckles and tapered receivers for tightening the window shut against the draught-seal.

Upgrading old windows

Does upgrading glazing necessitate replacing windows? Not unless they're rotten beyond repair. Super-thin double-glazing is made for upgrading old frames. Or you can fit stepped double-glazing into existing rebates and add beading. (For sash windows, however, this brings complications, as the additional glass weight requires heavier – so longer – counterweights. Also, to support this extra weight, side-hung casements may need additional hinges near the top.[10])

A simpler, albeit more cumbersome, alternative is to fit secondary frames and glass. This secondary unit should be double-glazed as the deep airspace (which is good for reducing noise transmission) allows convective air circulation, which compromises thermal insulation. Even if you'll never open these windows, hinge them for cleaning, and tighten them against the draught-seal (e.g. with tapered wooden turnbuckles). Sprinkle desiccant (e.g. silica gel, anhydrous calcium sulphate) into the inter-space to absorb condensation – or just leave the outer windows draughty, so air will dry it out. If there's cold bridging at the window reveals, secondary glazing (and insulation on the reveals, which further reduces noise) overcomes this.

Replacing windows

If you choose to replace windows, what should they be made of? uPVC windows are the cheapest. They're draught-proof, rot-proof and nowadays most kinds are designed with thermal-break sections, so they lose little heat through the frame. Some have integral roller blinds (which insulate like another layer of glass). PVC is (mostly) made from abundant raw materials (seawater); and many firms make bespoke windows to fit all shapes and sizes of openings. So far, so good. However, PVC is polluting to manufacture, and post-manufacture off-gassing raises health concerns (see: **Chapter 13: What we breathe**, page 151) – which their draught-proofing efficiency exacerbates. Longevity-wise, the fact that PVC is rot-proof doesn't mean it can't crack or break – which (with age, ultraviolet exposure, stress or damage) it can. Aesthetically, solid plastic products have no sensory appeal beyond the visual (see: **Chapter 14: Sensory nutrition**, page 170); and uPVC windows and doors compromise older buildings' appearance and authenticity (see: **Chapter 14: Soul/spirit factors**, page 173).

Fibre-reinforced plastic windows and doors are tougher than uPVC. Some have an embossed grain to look like wood. Apart from their greater durability, different gas-emissions and (patently artificial) textures, these share all the shortcomings of uPVC.

Aluminium and steel windows are also rot-proof. Modern ones are also made with thermal-break sections to minimise heat loss. The strength of folded metal allows frames and mullions to be very slim and the window panes they support to be very large, maximising both view and light. This clean spacious aesthetic comes at an embodied-energy price, though. Aluminium and steel are polluting to manufacture, and aluminium is particularly energy intensive (see: **Chapter 17: Materials we build with**, page 194); although, being expensive, it's widely recycled.

Wooden windows have a reputation for rotting. This, however, is mostly a characteristic of mid- to late-twentieth-century windows. Many Georgian windows are still sound, over 200 years after they were made. Good-quality modern timber windows can outlast PVC. Most softwoods need painting, oiling or protective stains against rain and ultraviolet, but some hardwoods need no treatment. As wood is hygroscopic, it swells when wet and shrinks when dry. This used to cause draughts, but modern water-repellents, EPDM draught-seals and dimensionally stable engineered timber overcome this. Wood has some thermal insulation value, but not enough for seriously low-energy houses. To minimise heat loss, therefore, low-energy windows are made of engineered timber: an insulating core (e.g. balsa) is sandwiched between durable outer faces (hardwood or softwood according to budget).

The most durable external face of all isn't timber, but an aluminium shield. This composite construction – aluminium-faced engineered timber – gives the most thermally efficient, long-lasting, low-maintenance and attractive windows money can buy. Unfortunately, however, it does take money to buy.

This makes a range from unhealthy to non-toxic, short-lived to durable, high to low maintenance, inexpensive to expensive, and industrial to 'natural'. Which you choose depends on your priorities.

Keypoints

- Replace single-glazing with, at least, low-E double-glazing; or low-E triple-glazing if circumstances permit.
- Double- and triple-glazing units are much heavier (and thicker) than single glass.
- Window material affects longevity, dimensional stability, indoor air quality and appearance.

Choices

Options	Decision-making factors
New windows, stepped double-glazing or secondary units?	Cost; appearance; sound-proofing; weight; cold bridge issues.
Replacement window material?	Toxicity; durability; maintenance; price; sensory and associative aesthetic; visual and historical context.

Insulating existing ground floors

Insulating old ground floors involves more work and usually much more disruption than most other energy conservation measures. Although techniques and mess vary with their construction, in all cases, you'll be working on a floor that you also need to stand (or kneel) on.

Solid floors

Many old ground floors are solid. The oldest ones are likely to be tiles over sand, but most are on concrete. To insulate solid floors you can do one of the following:

- Remove and replace the tiles, after adding insulation (and a damp-proof coating, if required).
- Remove the whole construction and lay a new slab and floor over insulation and (if the slab is concrete) a damp-proof membrane.
- If the whole building is used to being slightly damp – as many old ones are – remove the existing slab and replace with limecrete over granular insulation so everything can breathe (see: **Chapter 10: Warm ground floors**, page 111).
- Insulate over the existing floor and add a new surface.

Slab replacement is usually the best method, but is the most work and keeps the indoor air damp for longer while the slab dries out. (To break up a concrete slab, hire a pneumatic chisel to cut it free of walls and let you get a 5' (1.5m) bar (extended if required by a length of scaffold-tube) under one edge. Pry up one corner and kick in a small stone to hold it off the ground (a millimetre is enough), then hit it in the middle with a sledgehammer. *Kick* the stone (with steel-toe-capped boots): don't use your fingers unless you can do without them. This isn't hard work – unless the sledgehammer hits the ceiling, making *two* jobs for you now to do. Even easier: let a builder do it (see: **Chapter 16: Sequence of action, and by whom?** page 219).

Insulating *over* floors (or floor slabs) raises them, thus lowering ceilings, windows and, more critically, door-ways. Besides increasing head-bump risk, panel doors can look unbalanced if shortened much more than 75mm (3"). Height reduction doesn't affect sliding doors: just raise the track (head-bump risk, however, remains). Plywood doors are easy to cut to fit re-sized openings but, as they're minimally structured (it's the hexagonal paper core that keeps the plywood faces apart), you'll need to slip in a new internal frame member. I prefer, however, not to use them, as internal-grade plywood emits formaldehyde (see: **Chapter 13: What we breathe**, page 151). It's also easy to cut down boarded (ledged and braced; not *framed*, ledged and braced) doors. If you can't buy these in the width you need, they're easy to make.

These door-shortening implications often limit insulation thickness. Options for suitable, relatively thin over-floor or -slab insulation and finishes include:

- Vacuum panels. With proprietary 7-10mm tongue-and-groove floorboards and two thin foam levelling sheets,[11] these give a total thickness of around 40mm (1⅝"). As vacuum panels are very expensive but easily punctured by grit underneath them, they should be laid by a meticulous – or closely supervised – floor layer.
- Polyisocyanate board (K-value: 0.02). This is much cheaper – but gives less insulation. With 41-50mm

Panel doors shortened much more than 75mm (3") look unbalanced.

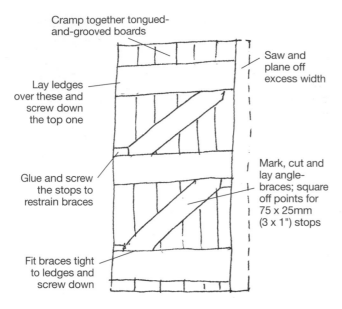

Cramp together tongued-
and-grooved boards

Saw and
plane off
excess width

Lay ledges
over these and
screw down
the top one

Glue and screw
the stops to
restrain braces

Mark, cut and
lay angle-
braces; square
off points for
75 x 25mm
(3 x 1") stops

Fit braces tight
to ledges and
screw down

Boarded doors are easy to make.

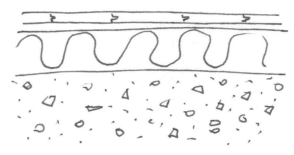

Insulating existing floors: 41-75mm insulation board, and proprietary 7-10mm tongue-and-groove floorboards over a thin foam levelling sheet.

polyisocyanate, underlay, then 15mm bamboo flooring, this totals approximately 60-70mm (2½-2¾").

- Wood-fibre insulation board (K-value: 0.039), with finishes as above. Although somewhat less insulating (unless you use a thicker board), this circumvents all potential chemical hazards. Although less energy efficient, it's healthier.

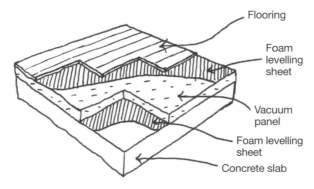

Flooring

Foam
levelling
sheet

Vacuum
panel

Foam levelling
sheet

Concrete slab

**Insulating existing floors: vacuum panels.
(Courtesy of Kevothermal Ltd)**

Suspended floors

Although solid floors are common in newer buildings and old farmhouses, most nineteenth-century townhouses (and many twentieth-century ones) have suspended floors. With these, you mustn't do anything that restricts their underfloor ventilation. Indeed, you need to check that airbricks and grills haven't become blocked. This is a common cause of dry rot.

If door height isn't problematical, you can lay insulation and a new finish over the existing floor – as for solid floors. Otherwise, you'll need to remove the floorboards and insert insulation between the joists – as for new suspended floors (see: **Chapter 10: Warm ground floors**, page 111). (For floorboard removal, it's sometimes easier to punch nails through than pull them out; then prise up boards from below.)

In-floor insulation is simple technically but, as you don't have a floor to stand on while doing it, more disruptive than over-floor insulation. If you can't face the disruption, you should at least draught-proof suspended floors. (Few old floorboards are tongued-and-grooved; most are straight-edged, which shrink, leaving gaps. Apart from the heat loss, cold ankles feel miserable.) To draught-proof floors, punch down all nail heads so they're below the surface, fill these and all gaps with papier-mâché or cellulose filler (e.g. Polyfilla), then sand, linseed-oil and wax the floor. Or lay hardboard on top, tape the joints and cover with a sheet finish (e.g. linoleum), cork tiles (warm to touch but not very hard wearing) or (harder-wearing) bamboo.

For all kinds of ground floor, removing (at least) the floor surface minimises the ceiling and door height reduction due to insulation. However, it's disruptive. Indeed, if you're living in the house, whatever way you insulate floors causes disruption. Whether you are or aren't already living in the house is likely to influence the choice between minimum disruption and minimum heat loss. Bear in mind, though, that disruption is only once, whereas heat loss is for the house's lifetime.

Keypoints

- Insulating over floors raises them, lowering ceilings, windows and doorways.
- Alternatively, remove and replace the whole floor construction – very disruptive but a better (warmer and drier) job.
- Suspended ground floors can be insulated between the joists – but this requires removing and re-laying floorboards, so is very disruptive.

Choices

Options	Decision-making factors
Remove and replace whole solid floor or insulate over it and add new surface?	Existing floor condition; disruption; drying-out time; effort; headroom.
Excellent (expensive), adequate (mid-price) or poor (thin and inexpensive) insulation on solid floors?	Price; headroom; door-shortening implications; energy/comfort; health (off-gassing); thermal perimeter elsewhere (e.g. extension, skirt, underground external insulation).
Insulation material and technique choice for suspended floors?	Airborne fibres; price; access to underside of joists; depth available for insulation; disruption; rat-proofing.

Old chimneys: problems or assets?

Most old houses come with old chimneys. Until the 1970s, most houses were heated by open fires, but we no longer heat with coal, so chimneys are redundant. All take up space. Many leak. If you'll inhabit the roof-space, chimneys make cold bridges where passing through the roof. Are they any use?

Pre-Second World War houses typically had a fireplace in every room. One chimney flue is useful if you want to install a wood-burning stove, but four? Six? Because of their height, flues draw air out of rooms – so foster incoming draughts. Additionally, most old chimneys are slightly damp, as their damp-proof-course (to stop rain wetting brickwork below) is rarely stepped to overlap their flashings. Traditionally, this moist brickwork dried out within the attic so rooms below were dry. If you use this space, however, you want it dry. If there's any damp, this makes a case for demolishing the chimney above roof level (or, more expensively, rebuilding it with a lead tray lapping the flashings). This, however, denies the option of future use and can compromise the house's character. It's wise, therefore, to at least cap unused flues against rain (retaining extract ventilation under the caps) and to incorporate an adjustable grill when walling up a fireplace.

The other issue is space. Chimney breasts consume expensive floor space. It's tempting, therefore, to demolish them. This is possible, but rarely advisable. Although instability and damp require some chimneys to be removed above roof level, complete demolition is expensive and disruptive. Moreover, chimney *breasts* add thermal mass and generally stabilise walls, which can slightly improve room-to-room sound insulation.

Additionally, flues can be useful. If a chimney abuts a bathroom, kitchen or larder, you can use it as an extract vent, and seal the fireplace. You can also use chimneys' powerful ability to extract air as part of a summer cooling system. Extending the flue by replacing the chimney pot with a longer, black-painted (so solar-warmed) lightweight pipe increases its draw. (How high this can extend depends on stability: it may need to be braced or guyed.) As you want neither draught nor cooling in winter, you'll need to fit a hinged or butterfly shutter or an adjustable grill to open or totally seal off airflow. In summer, you might not mind where extract vents are: any will do. But in winter, you want fresh-air inlets in habitable rooms and outlet vents only in moisture / odour-producing rooms. If you don't need flues as air extracts, you can block them with inflatable 'balloons' – unlike filling them with concrete, this retains the option of using them at a later date.

In short, although unused chimneys uselessly take up space, you might one day need them for heating stoves or cooling ducts, and they have other uses too – so think twice before removing them. Then think twice again.

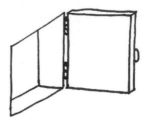

Butterfly and hinged shutters to control ventilation extract.

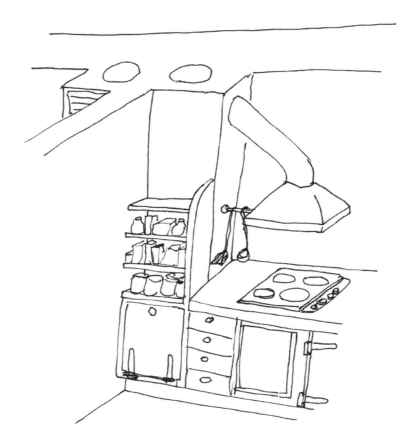

A chimney abutting a bathroom, kitchen or larder can house extract vents.

Keypoints

- Retain the option of using unused flues for wood stoves / gas boilers.
- Cap them against rain.
- Ventilate unused flues to keep chimneys dry.
- Utilise flues for ventilation of wetrooms, kitchens and larders; and for summer cooling.
- Use a solar-warmed extension pipe to increase draw.

Choices

Options	Decision-making factors
Retain or demolish unused chimneys?	Thermal mass; structural stability (above and below roof); floor space need; leaks at flashings; damp transmission.
Seal or keep flues open?	Possible future need for wood stove / gas boiler flues; dampness; ventilation potential for wetrooms/kitchens/larders and for summer cooling.

Resources

Information

Cotterell, J. and Dadeby, A. (2012) *PassivHaus Handbook*, Green Books, Cambridge.
Fairweather, L. and Sliwa, J. (1997) *AJ Metric Handbook*, Architectural Press, London.
www2.buildinggreen.com
www.deac.co.uk/advice-from-deac/insulation-houses

Less common products

Insulation stilts

B&Q: *www.diy.com*
www.ijoistflooring.co.uk/images/techguideflloorapp.pdf
www.loftzone.co.uk/storefloor.html
www.screwfix.com

Bamboo flooring

http://simplybamboo.co.uk/products
Borate rods, pastes and timber treatment:
www.boron.org.uk
Glass: *www.slimliteglass.co.uk*
Wood-fibre insulation board: *www.naturalinsulations.co.uk/index.php?location=SteicoRigid*

1. *www.deac.co.uk/advice-from-deac/insulation-houses* (accessed: 12 December 2013).
2. *www.thermalcalconline.com* (accessed: 16 December 2013).
3. This may not be necessary, but it's a wise precaution that (if you use bottle-top spacers) costs nothing.
4. If using battens thinner than 25mm or spanning over 400mm, you'll need to screw plasterboard. They're too bouncy to easily nail to.
5. Sometimes, they're only 1½" – not enough!
6. *www2.buildinggreen.com* (accessed: 8 April 2014).
7. The minimum distance depends on the partition's thermal conductivity, indoor humidity, indoor–outdoor temperature differential and wind-chill. If in doubt, extend the insulation.
8. Paticas, H. and Clarke, A. (2014) London upgrade future-proofs historic building, *Passive House+*, 2014, Issue 8.
9. If using intumescent paint, ensure there's lots of ventilation as the solvent fumes aren't good to breathe.
10. Put these hinges two-thirds to three-quarters of the way up.
11. Some kinds of flooring, however, are so heavily synthetic-varnished that they're little better than plastic.

CHAPTER TWELVE

Keeping cool

Stack-ventilation (page 140)

Evaporation (page 138)

Thermal mass (page 138)

Shading:
Fixed
Adjustable
(page 134)

Foliage shade and transpiration (page 146)

Air movement (page 138)
Cross-ventilation (page 140)
Cool night air (page 138)

Cooling basics

Most eco-design concentrates on keeping homes warm. However, if cooling isn't also considered, low-energy homes can be *too* hot in heatwaves. Many energy conservation improvements ignore this – so, even in normal summers, some eco-homes are too hot. And, with global warming, we must expect increasingly frequent and insufferably intense heatwaves. As thermal resilience declines with age, the elderly – which we all hope to one day become – are especially vulnerable. Moreover, the disabled can't easily take remedial action (e.g. holding wrists under cold water, wetting hair, draping shoulders and spine with damp cloths). Tragically, fuel poverty has familiarised us with 'cold deaths', but 'heat deaths' are unfamiliarly new in mild climates (e.g. Britain, northern France). With climate change, their numbers are bound to increase – and you don't want to be one of them. It's important, therefore, that, besides being warm in winter, homes must *also* be comfortably cool in summer. More exactly, how can *we* be cool in them? (Not quite the same thing.) How can this be achieved without resorting to energy-expensive air-conditioning?

In air at blood temperature, the only ways our bodies can cool are by conduction or radiation to cool things (rare, as most are at air temperature) or by evaporation. Even at around 30°C (86°F), it's difficult to lose the heat produced by metabolism fast enough. As a result, we sweat profusely (for evaporative cooling) and feel limp (to minimise metabolic heat production).

For our environment to cool *us*, we need cool air – or, if not cool, at least dry and moving, so sweat can evaporate. For cool air, we need high rooms, so the warmest air is well above head-level. For air movement, these allow ceiling-mounted fans to be mounted above decapitation-level.[1] We also need windows or other ventilation – and space. For cool surfaces around us to radiate, and sometimes conduct body heat, to, we need the building fabric to remain cool, not warm up – as it tends to in warm weather. Consequently, the building form should be compact to minimise warming surface in daytime: the exact opposite of both that needed to cool it at night and for cross-ventilation air movement. Baruch Giovani, therefore, proposes insulated shutters on indented porches so buildings minimise perimeter area (hence heat-gain surface) when the shutters are closed in daytime, and maximise it for cooling when they're open at night.[2]

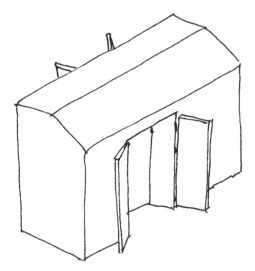

Flexible building form: compact by day to minimise heat-absorbing surface area, extended at night to maximise cooling surface area.

We must also *cool* the building itself. For this, we can use two principles: warm air rises (for extract ventilation) and cool air drops; and evaporation cools. The combination of warm and evaporation-cooled air propels air movement.[3] Both are driven by the power of the sun. So, the hotter the weather, the more powerful its cooling effect. Nature does this in several ways. Heat propels evaporation, producing rainforests, which give deep shade. In drier regions, summer heat bleaches ground so it reflects solar rays. At the other extreme, snow reduces cooling, both by insulating ground beneath it and by reducing heat-radiation to space during winter's long nights. Similarly, by utilising the forces of nature, climate-sensitive design can moderate thermal extremes. Indeed, in most climates, thermal *comfort* is attainable under almost all circumstances, merely by the passive functioning of buildings – if they're appropriately designed.

Keypoints

- Metabolism produces heat so our bodies must cool.
- In warm air, we cool mostly by radiation to cool surfaces and sweat evaporation, speeded by air movement.
- Room height keeps warm air high. Building height maximises stack-effect air movement.
- Air can be cooled by water evaporation.
- Compact forms minimise warming surface in daytime. Extended forms maximise cooling surface at night.
- It's important to cool the interior building fabric so heat doesn't build up.

Keeping heat out

Everyone knows how important it is that buildings can keep heat in. But unless they can keep heat *out*, they'll be uninhabitable in heat waves, especially in muggy, still-air conditions. The most important heat to keep out is sunlight. Hot air is hot, but convection from sun-warmed surfaces makes it hotter – especially indoors. For both windows and outdoor sitting places, this makes shading important. The need to shade from direct sunlight is obvious, but there's also reflected light (e.g. from glazed façades – which tree screens can obstruct) and re-radiated heat (e.g. from paving – which green groundcover outside windows can eliminate).

Overhead awnings (especially synthetic fabric ones without vents at the top) heat up and radiate heat on to the head – so they can be more exhausting than full sunlight to be under. Similarly, sun heats up blinds (although less so if they're made of natural ex-living materials, as, when alive, these kept living organisms at life-maintaining temperatures). Consequently, internal blinds transmit 80-90% of this heat indoors.[4] External blinds just warm the outdoor air. Exposure to weather, however, puts extra demands on their construction and/or shortens their life. In many ways, external slatted shutters, traditional in many countries, are the best option. (That's why they're traditional.)

Adjustable slats let you choose how much light and view you want, and (if it's not too hot) allow air in. External louvres use similar principles: spaced apart, they needn't obstruct view or airflow; and, if adjustable, they can be optimised for sun angle, although not sun direction (i.e. time of day). Brise-soleil, either fixed or adjustable, deal with time of day, but not sun inclination. Shutters do both. If side-hinged, you can open one side or the other to shade morning or evening sun.

Deep roof overhangs, awnings or even solar panels can ensure summer noon shade for south windows. But in afternoons, the sun is low – and the heat most intense. Afternoon shade, therefore needs to be from the side: e.g. by trees, bulky shrubs, vines on trellises or side-hung brise-soleil, shades or shutters.

Walls also need to be shaded. Vine shade on walls can reduce summer solar heat gain by up to 95%.[5] Remember, though, that vine stems grow big enough

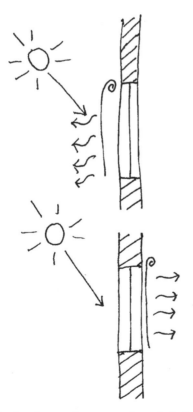

Outdoor shades disperse heat to outdoor air; indoor ones, to room air.

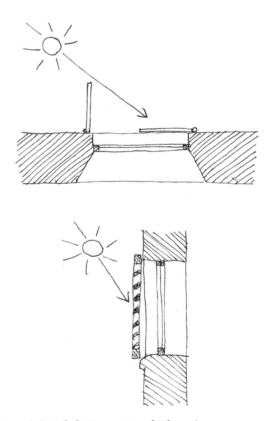

External slatted shutters: many shade options.

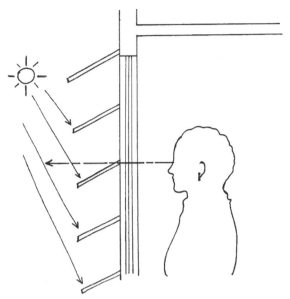

Louvres shade from high-angle sun without obstructing view.

for intruders to climb on, and smaller pests can use smaller shoots to access unscreened open windows. Living (green) walls both stop heat transmission indoors and cool outdoor air. Those requiring waterproof walls, however, compromise breathability.

Overheating is a summer problem; and in summer, the sun is high, so shines *down* on to roofs. Consequently, tree shade can be worth 100 air-conditioner-unit hours/day.[6] Vine shade is quicker to establish and also eliminates bough-fall risks. Carpeting roofs with Virginia Creeper can reduce temperature over 11°C (23°F), thus reducing cooling costs by 73%[7] (see: **Chapter 3: Choosing what should be where in the garden**, page 23; **Chapter 15: Burglar-proofing**, page 180). Green roofs are another possibility. In winter, their soil thickness and leaf-entrapped air keeps roofs around 4.5°C (8°F) warmer and can halve heat loss due to wind. In summer, they both insulate against heat gain and transpire and evaporate water, so keep roofs an average of 15°C (27°F) cooler.[8] As they need irrigation in prolonged dry weather, this produces yet more cooling. (Occasional drought damage is rarely irreparable for green

Noon shade: roof overhangs, pergolas, awnings.

Afternoon shade: brise-soleil, vines on trellises, side-hung shades/shutters, trees/shrubs.

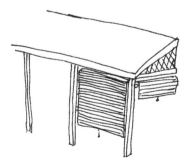

Canopy roofs with roll-down blinds give many shade options.

roofs because of the characteristics of the plants used: grass seeds before it dies, and some sedums (e.g. stonecrop) survive dry conditions.)

If they're in sunlight, most roofing materials (e.g. tiles) are dark enough to warm. Black asphalt and bitumen (common on older buildings, e.g. conventional flat roofs, bitumen-fibre corrugated sheets and shingles) make the worst kinds of roof for heating up in summer, so need a reflective surface covering (e.g. a white or pale surface) and/or lining. White roofs are *much* cooler than conventional mid-tone ones: in Florida, they can reduce cooling bills by 40%.[9] However, they're often visually unacceptable: conspicuously non-traditional and horrid when dirty.

Eventually, the under-roof insulation warms up too. If possible, therefore, incorporate at least one reflective under-layer. Also, ventilate the airspace between hot surface and insulation, allowing generous outlets (e.g. whole ridge, chimney or rotating cowl).

It's an oversimplification to regard overheating as just a summer problem. More specifically, it's a summer *afternoon* problem. This makes the time it takes for heat to travel through the roof relevant. Thin and conductive roofs radiate heat into rooms while the evening air is still warm, making sleep

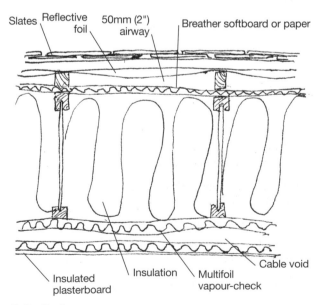

Slates   Reflective foil   50mm (2") airway   Breather softboard or paper

Insulated plasterboard   Insulation   Multifoil vapour-check   Cable void

Reflective layers.

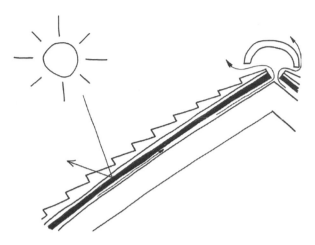

Roof cooling: generous ventilation, reflective layer(s).

Keypoints

- Shade windows and outdoor sitting places.
- External shades disperse heat to outdoor air; indoor ones, to room air.
- Plastic and metal shades heat up, then radiate heat. Natural materials do so much less.
- Deep roof overhangs can provide summer noon shade. But in afternoons, shade is needed from the side.
- Roofs should incorporate a reflective layer (and/or be white). Generous under-roof ventilation prevents heat build-up.
- Climbing vines can shade walls; trees (or vines on frameworks) shade roofs.

impossible. West walls warm up and windows admit sun just when roof heat arrives indoors. Green and brown roofs have significant thermal mass, so can both store night-time coolth and slow heat transfer till the cool morning hours.

As most summer overheating is due to roof construction, there are two essential rules: ventilate generously and incorporate one (or several) reflective layers. Thermal mass also helps – but only if it delays heat transfer long enough. Shade makes a huge difference.

Choices

Options	Decision-making factors
White roof coatings or reflective foil within roof?	Visual impact; roof construction, materials and thickness; cleaning access/frequency.
Internal or external blinds?	Maintenance access; longevity; wind/rain damage; cooling efficacy.
Noon shade or afternoon shade?	Which gain more heat: roof or south-west / west wall(s)?
Fixed roof overhangs, awnings, adjustable canopies, brise-soleil or foliage shade?	Predictability of summer weather; access/ease for adjustment, maintenance, pruning.
Vine-covered walls and/or roofs, green walls and/or roofs?	Planting bed possibilities; likelihood of summer drought; roof irrigation feasibility; rodent/pest/intruder access to windows.
Sucker, rooting-tendril or twining vines?	Wall construction/materials; ease of fixing wires, frameworks; access to prune; risk of clogging gutters.
Watered roof?	Feasibility; slope; water availability.
Overhanging tree shade?	Bough-fall risk; access to prune; intruder access to windows; chimney draw.

Keeping cool indoors

Like being warm, being cool depends on a number of interacting factors: air temperature, humidity and movement; radiation to cool surfaces (and outer space); the extent to which air cools or warms surroundings; and the capacity of surroundings to cool or warm air. So should we focus on cooling the body, cooling the air or cooling the fabric of the building? This affects design, choice of materials and, particularly, location and design of windows. Which we choose, however, depends on circumstance.

For cooling building's interior fabric, there are basically three passive and three active ways: evaporation, natural air movement, and cool night air impoundment; and fan-driven air movement, pumping cold fluid, and refrigeration-cooling air (air-conditioning).

Evaporative cooling is a long-established Middle Eastern technique, of which there are modern variations (e.g. Arizona Cool Towers, add-on evaporative cooler units). Air movement (e.g. by cross- and stack-ventilation) doesn't cool when air is hot in daytime, but without any, buildings add incidental gains (e.g. from people, cooking, lighting) to air temperature, so warm up more each day. Air movement is most effective at cooling during the night time, when air is cool. Indeed, if heat is extreme, windows must be *shut* by day.

Heavy thermal mass stabilises temperature so, besides keeping houses warm in winter can cool them in summer.[10] For solar heating in winter, thermal mass needs to be in the right *place*. For cooling in summer, it needs to be cooled down at the right *time*: night time – as night air is cool. Cool air is heavier than warm air, so high-level openings, open at night but closed in daytime, can let night air form cool 'air ponds' indoors. This impounded cool-air reservoir cools the building's thermal mass and can last all day. Normally, plenty of night-time ventilation suffices for this. It requires openings proof against pests (including neighbours' cats) and intruders. This implies mosquito-screened (and grilled or carefully positioned) windows and perhaps also lockable grills on external doors, so they can be left open at night.

(Remember to hang the key in an obvious place in case of fire – but out of strangers' sight and grab-stick reach!) Underground homes, traditional in some hot arid regions, combine cool-air drop with massive thermal mass to remain deliciously cool all year. (They're thermally stable in damp regions too, but must be waterproof against soil moisture and perhaps groundwater pressure; and, as water runs downhill, risk filling up if heavy rain gets in.)

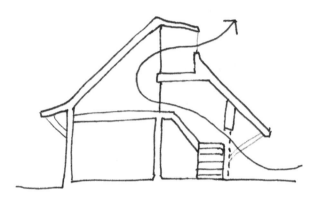

Inlet air from the cool side of the house, outlet by stack-ventilation, with lockable grill on door and mosquito-screening on all openings for secure and pest-proof 24-hour ventilation.

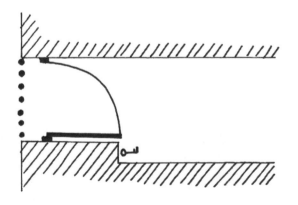

Key in case of fire.

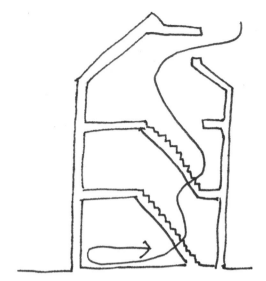

High-level openings for deep cool 'air ponds'. However much air circulates, the coolest air sinks, so can be trapped at night. With the thermal mass that it cools, this forms a reservoir of coolth for the day.

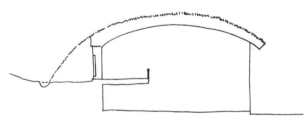

Undergound homes need well-drained entrances.

Nowadays, mechanical cooling is more common than passive methods. Mechanical systems act faster but are more elaborate, so there's more to go wrong. These include pumping cool liquid (either cold water – e.g. stream- or pond-cooled – or heat-exchange fluid from heat pumps[†] run in reverse) through the same radiators as are used for heating in winter or through piping embedded in thermal mass. Unlike winter heating, however, it's best to cool the fabric above us, so the air it cools drops down on to us. The most common form of mechanical cooling is air-conditioning. This, however, is electricity intensive, so is even more anti-eco than coal-burning open fireplaces. If you want this, look elsewhere for information.

Cooling buildings is only half the story. More important is cooling *us*. To us, motionless air feels dead; and motionless hot air, stifling. Air movement can be fan-driven, or be diverted, focused, enhanced and induced by building form. Fans use little electricity. They're 95% cheaper than air-conditioning to run, and 90% cheaper to install.[11] Natural air movement, though, is free – and still works in power-cuts. Air movement speeds evaporation, so we're cooler in a draught (e.g. between two open windows). But draughts aren't always comfortable (or healthy) and they blow papers around, so don't suit paper-based activities. Directing air currents (by keeping openings high or using bottom-hinged windows) to sweep the

Shade screen over entrance

Mid-twentieth-century underground home in central California (where summer temperatures frequently reach 114°F/46°C): cool night air drains to lower levels. From the ground, oranges can be picked from light-well trees. Note that there are no doors or windows. Although winter night air is chilly, the thermal mass of the earth keeps temperatures tolerable all year. Security is by dogs, not doors.

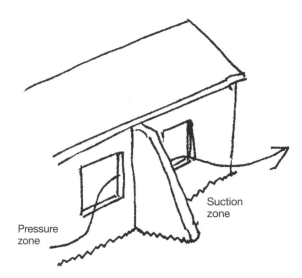

'Wingwall' hedge/wall.

Pressure zone

Suction zone

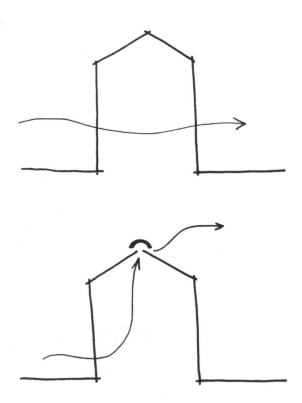

Cross-ventilation, stack-ventilation.

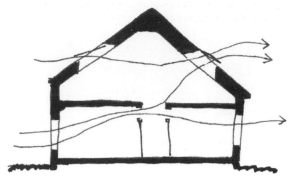

Whole-house ventilation combining both wind (or sun-shade pressure difference) and stack effect. Note the vent slots over internal doors (which should be closable for acoustic privacy).

ceiling flushes away the warmest room-air and can – if outdoor air isn't too hot – help cool the building fabric. Will cooling the building suffice, though, or should draughts also cool inhabitants? Will heat coincide with doldrums-airlessness, gusty winds, or thunder showers? The more ventilation options, the easier it is to optimise ventilation under any weather conditions. This means, for instance, adjustable openings on windward and leeward and shaded and sunny sides for cross-ventilation, and at high and low levels for stack-ventilation.

Any cooling involving evaporation relies on reasonably dry air; and any involving cool night air relies on reasonably clear night skies. In the humid tropics both are rare. Consequently, in humid climates, only air movement cools, and buildings should have *little* thermal mass, so they don't store heat.

What if the weather is sometimes hot and humid, sometimes hot and dry? The determining factor here is night-time temperature. If this usually drops enough in the small hours to adequately cool the interior thermal mass,[12] heavyweight construction can keep your home cool by day. If it doesn't, sleep on the roof or balcony when it's too warm – under an awning or lightweight rain-roof (and mosquito net) if necessary. For indoor cooling in humid conditions, air movement is the only (passive) option – after, of course, minimising all heat gains.

Cross-ventilation using windows to windward and leeward, and to shaded and sunny sides of the house.

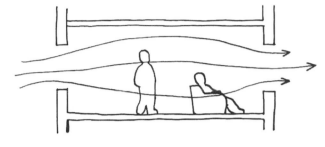

Draught-free cross-ventilation using hopper-type windows.

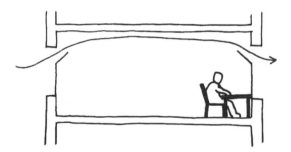

Cross-ventilation to cool us.

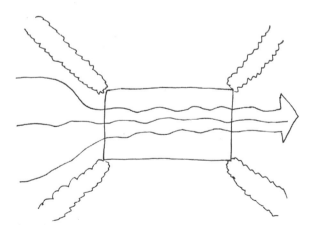

Hedges to deflect cooling breezes.

Even in still conditions, cross-ventilation and stack-ventilation can create some air movement. Additionally, diagonal hedges from building corners can help deflect faint breezes indoors – and also create wind-protected sitting places for spring and autumn. They also help tie buildings into place visually. If, however, wind blows along façades rather than on to them, it won't produce cross-ventilation. Nor can there be any cross-ventilation if we shut internal doors and don't want open vents above them (e.g. because we want acoustic privacy). In such cases, you can contrive airflow. With *two* windows in each room, and a short transverse 'wingwall' or hedge (preferably up to eaves) between them, breezes are deflected into one window and sucked out of the other. (In winter, wingwalls slow wind along the wall, reducing wind-chill.)

Within a house's lifetime, how hot might summers get? Predictions are that British summer temperatures will rise 6°C (12°F) from 2009 to 2050.[13] In heatwaves, this raises temperatures into the danger zone. Although prediction is an unreliable science, this is a warning we must take seriously. You don't want to build or buy a house that'll be uninhabitable before you've paid off your mortgage. This, therefore, requires thinking about summer cooling *as well as* about minimising winter heating.

Keypoints

- Thermal mass stabilises temperature.
- In hot weather, thermal mass must be cooled down at night using ventilation. Night-time ventilation requires intruder- and pest-proof openings.
- High-level openings admit night air to form 'cool-air ponds' which last all day.
- Cross-ventilation and 'stack-ventilation' maximise air movement potential.
- Hedges and 'wingwalls' can deflect faint breezes indoors.
- The more ventilation options, the easier can ventilation be optimised to suit conditions.

Choices

Options	Decision-making factors
Cool people, air or building fabric?	Draught nuisance or delight; night-time security; temperature/heat build-up; room-to-room airflow; options for window positioning and design.
Radiant cooling by exposure to cool surfaces?	How is fabric cooled: night air, piped cold water or evaporation?
Location of cooled thermal mass?	In ceilings to drop cool air? Or in floors to be cooled by ponded night air? Security issues if windows must be open at night.
Evaporative cooling?	Where? How? Is there sufficient water?
Cooling by air movement: wind, cross-ventilation, stack-ventilation or thermally induced?	Local climate; building/landscaping form.

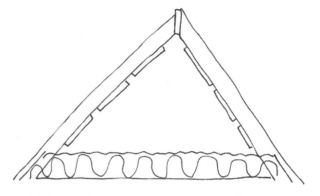

Multifoil stapled to rafters, with gaps for ventilation.

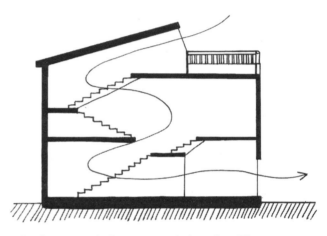

Roof terrace stair doorway as wind-catcher. If breeze direction is diagonal, the door can be outward-opening and hinged accordingly.

Keeping cool in an old building

Most eco-design focuses on minimising heat loss, but global warming makes cooling a serious issue even in mild climates like Britain. The UK Climate Impacts Programme predicts that, by 2021, lightweight buildings will need air-conditioning.[14]

Outside the humid tropics, it's rare for air to be unbearably hot all 24 hours, so thermal mass helps keep buildings tolerably cool (see: **Chapter 8: Thermal mass**, page 83; **Keeping cool indoors**, page 138). Old masonry buildings have significant thermal mass (see: **Chapter 11: Old chimneys: problems or assets?** page 129). Don't demolish this. Keep it. Wherever possible, keep it within the insulation perimeter. (Timber-framed buildings have little thermal mass

other than brick chimneys, but you can easily increase it by adding another layer of plasterboard.) Often, however, old masonry walls have been tidied up by 'dot and dab' dry-lining. Their surface is now plasterboard *decoupled* from the underlying thermal mass. Removing this plasterboard entails considerable disruption: mess, dust and a legacy of electrical cable and steel switch-boxes on the wall surface. If you decide to re-plaster, remember that your home will take time to dry out.[15] (Re-plastering isn't obligatory: sometimes exposed brickwork looks better, although the indoor faces of external walls should be repointed

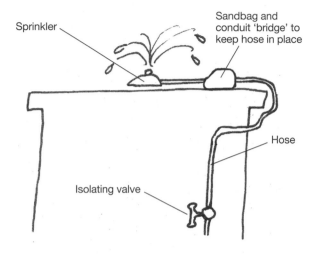

Flat roofs can be sprayed with water to cool them.

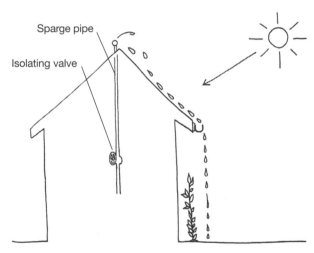

On sloping roofs, more water runs off before evaporating. Use this to irrigate shade vines.

against air leakage. Nonetheless, if you can face the disruption, you'll be more likely to have tolerable indoor temperatures in summer heat-waves, more stable warmth in winter – and no 'mystery' draughts (see: **Chapter 7: Airtight construction**, page 70).

Despite their high thermal mass, older buildings weren't designed for cooling: their top floors can become intolerably hot – too hot to sleep in. The first and most essential step, therefore, is to keep any additional heat – particularly sunlight – out (see: **Keeping heat out**, page 133). This includes vine-cladding roofs and shading west walls and windows (e.g. with veranda roofs and drop-down shades, vines, or trees); also perhaps retrofitting living walls. Resurface any black bitumen roofs with a reflective coating.

The second step is to keep solar heat from penetrating the insulation perimeter. Conventional insulation materials against cold aren't very good at this. Few of us feel cooler wearing sweaters on a hot day. This may require adding a reflective layer (e.g. multifoil) spaced below the ceiling, then more spacers, then a new ceiling. (You don't need the old ceiling but removing it makes a huge mess.) For cold roofs, thin (so cheap) multifoil can be stapled to the rafters, leaving gaps for ventilation. You might even need to do both.

Induced air movement is the third step. Breeze-catching additions (e.g. scoops, wingwalls: perhaps doubling as sheds, log-stores, etc.) can divert breezes indoors. Wind blowing along narrow 'breezeways' between two buildings (or walls) can drag air out of windows. Solar chimneys can accelerate air extraction so cool air can be drawn in from shaded or earth-cooled sources (see: **Keeping cool indoors**, page 138).

Then comes active cooling, by water evaporation. You can spray or mist water on to flat roofs (e.g. with misters, sprinklers, spurge-pipes) to cool them down. This also works on sloping roofs although more water runs off before evaporating, so the spray- or spurge-pipe should be along the ridge; and the spray periods should be briefer and run-off water diverted to irrigate shade vines.

Air-conditioning is only the last – and ecologically the least desirable – step.

Keypoints

- Add a reflective surface to black asphalt/bitumen roofs.
- Consider occasionally watering roofs.
- Maximise indoor airflow: use solar-boosted extracts and draw replacement air from cool sources.
- Retain – and expose – all possible thermal mass within the insulation perimeter.

Choices

Options	Decision-making factors
Vine shade, tree shade, fixed shades, solar panel shade, awnings or adjustable shading devices?	Maintenance access; noon or afternoon shade; weather and season predictability; root-bed possibilities.
Reflective surfaces, multifoils or conventional insulation?	Roof surface, ceiling or inter-space access; location and extent of unshaded exposure.
Cross-ventilation, stack-ventilation or breeze-catchers?	Breezes when needed; height of building; predictability of breeze direction; vent potential of chimneys; breeze-deflector / wingwall / hedge possibilities

Keeping cool outdoors

Indoors in hot weather, we feel trapped, but outdoors may be too hot for comfort, even in the shade. How can we cool outdoor spaces?

Shade, of course, is fundamental, but shade trees shouldn't restrict air movement. High-level foliage lets air move at person level and hedges can channel breezes. Additionally, just as sweat evaporation cools our bodies, evaporation from water-features cools the air. Mister sprays do so even more. (These must use clean or sterilised water, as Legionella is transmitted by breathing fine droplets.[16]) Transpiration from green-

ery (e.g. lawns, trees, shrubs, 'living walls', green roofs), and evaporation from watered sand-bedded paving, also cools the air. Consequently, gardens typically lower urban temperatures 1-2°C (2-4°F) and, locally, another 2°C where tree shaded.[17] With evaporation from sprinkler irrigation, lawns (or green roofs) can cool air 6-8°C (11-14°F).[51]

Water-features contribute cooling (also see: **Chapter 17: Things we affect**, page 202) for how to overcome associated problems). The sound of trickling water brings refreshment and soul balm, and so adds psychological cooling to physical cooling. Its glitter in sunlight is enlivening, whatever the temperature. Water-features can be wind-pumped or photovoltaic powered – so they run faster when it's windy or hot. Variable flow suits trickles and sheets of water, but not flowforms or fountains, which need a constant, precisely adjusted flow-rate. You also need to protect piping, valves and pumps from freezing in winter; and if overflow or splash can reach paths or roads, turn off water to prevent ice hazards. All open water evaporates in hot weather. It evaporates faster if it's splashing, spraying or warmed. As mains water may be in short supply, it's more ecologically responsible – and free from use-restrictions – to use rainwater or (shallow) groundwater for evaporative cooling and amenity features.

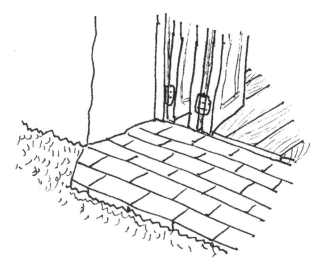

Watered sand-bedded paving can cool air 2-5°C (4-9°F) all day.[19]

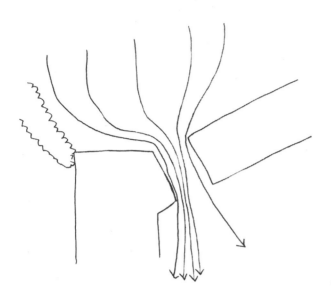

Draughts concentrated, channelled or induced between buildings.

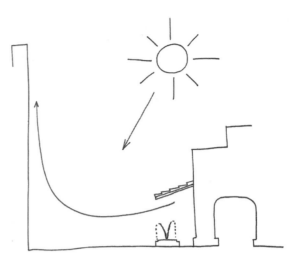

Central Asia:[20] temperature differential draws fountain-cooled air across the courtyard to rise up the sun-warmed wall.

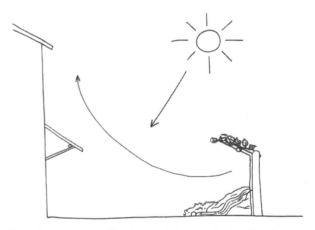

A modern version with photovoltaic-panelled roof and water-feature.

In hot weather, greenery is balm for the soul, but greenery needs water. If there's not enough, use thick mulches, only sub-surface irrigation and 'forest-garden' shade. Or, better, take a 'xeriscape' approach to landscaping: limiting plants to those that need little water (e.g. olives and many Mediterranean and arid-land species). There are also architectural means of cooling outdoor places. These include concentrating, channelling or inducing draughts between buildings (variations on techniques for moving air into, through and out of buildings; see: **Keeping cool in an old building**, page 142) and cool-air impoundment (see: **Chpater 3: Choosing what should be where in the garden**, page 23).

One traditional outdoor-cooling model is the court-yard house. Courtyards pond cool night air and mas-sive construction (e.g. masonry, adobe) stores this coolness. At least one wall, usually colonnaded, offers shade. There's usually a fountain and pool; and a tree and/or plants, which are generously watered at night, so paving is damp – adding evaporative cooling to radiant cooling from the thermal mass.

A Central Asian variation on courtyard-and-pool cool-ing is the addition of a tall, dark south-facing wall, which heats up in the sun; the fountain is on the shady south side. The ensuing temperature differential draws fountain-cooled air across the courtyard to rise up the sun-warmed wall. A modern version could use a pho-tovoltaic-panelled roof (which warms up) instead of the solar wall; and a water-feature, cascade, well-watered green-wall, water-wall or water-sluiced shade roof instead of the fountain.

There are many ways to use differential warming to create airflow. It is, after all, the way nature creates winds, from destructive hurricanes to door-slamming draughts between sunny and shaded sides of buildings. Nature's principles *always* work – so we might as well use them for thermal comfort.

Keypoints

- Shade is fundamental, but shade trees shouldn't restrict air movement.
- Concentrate and channel breezes or induce draughts between buildings.
- Courtyards pond cool night air and thermally massive construction stores this coolness.
- Use temperature differential to draw water-feature-cooled air across courtyards.
- The sound of trickling water adds refreshment and soul balm.
- If there's any risk of breathing droplets, water must be Legionella-free.
- Evaporation from water-features, greenery and watered sand-bedded paving cools air.

Keeping homes warm and cool

Many places suffer from both bitter winters and stifling summers. In continental climates, seasons tend to be predictable, but progression can be rapid. This tends to mean exclusively indoor living followed by mostly outdoor living – so it needs to be easy to carry food out from the kitchen; or even have a skeleton outdoor kitchen (grill, sink, evaporation-cooled chamber). Homes in such climates also require heavy insulation against cold and maximised exposure to sun, then maximum shade (indoor and out) and reflection of solar heat.

These contrasting requirements may sound irreconcilable. Deciduous foliage, for instance, is seasonally matched to sun and shade needs, but opposite to wind-shelter and cooling-breeze requirements. But summer and winter use-patterns – outdoor and indoor living – differ radically. Hence low-level pruned openings can admit breezes unimpeded at person-level *outdoors* for summer, but evergreen foliage or dense twigs above this shelter the *house* in winter. Sails and breeze-deflectors also help. Although seasonal winds are somewhat predictable, it's best if these are adjust-

Choices

Options	Decision-making factors
High or low shade trees?	Accessible fruit/nuts; winter sun for solar collectors / windows / sun-room; airflow at person level; noon or afternoon shade?
Hedge- or building-channelled draughts, or night-cooled courtyard?	Daytime breezes; plot size; social layout indoors; house design.
Watered paving or lawn?	Garden size; location of outdoor dining; foot-traffic concentration.
Water for psychological cooling?	Child safety; location for view, reflection, audibility; algae; mosquitoes; leaf fall; pump/filter maintenance; availability of water.
Misters? Sprays? Water-features?	Maintenance; health issues; assured availability of water.
Water features: variable with weather or constant, flow-rate?	Type; summer weather; night-time operation; icing in winter; assured availability of water.

Warm in cold winters, cool in hot summer.

Summer: +30oC (86ºF)

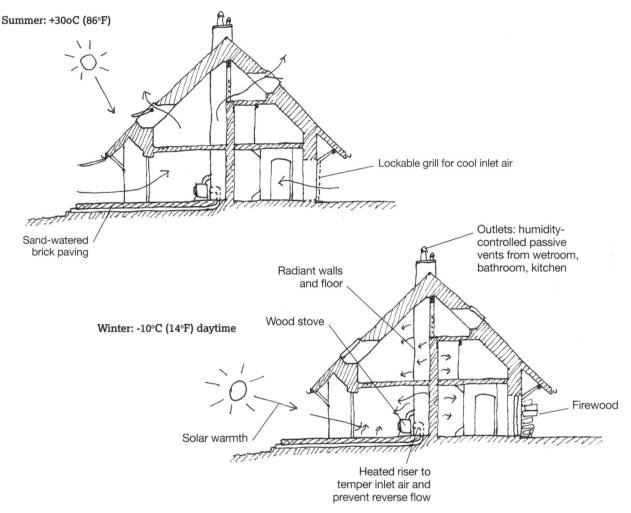

Lockable grill for cool inlet air

Sand-watered
brick paving

Outlets: humidity-
controlled passive
vents from wetroom,
bathroom, kitchen

Radiant walls
and floor

Wood stove

Winter: -10ºC (14ºF) daytime

Firewood

Solar warmth

Heated riser to
temper inlet air and
prevent reverse flow

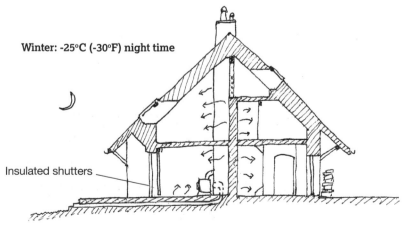

Winter: -25ºC (-30ºF) night time

Insulated shutters

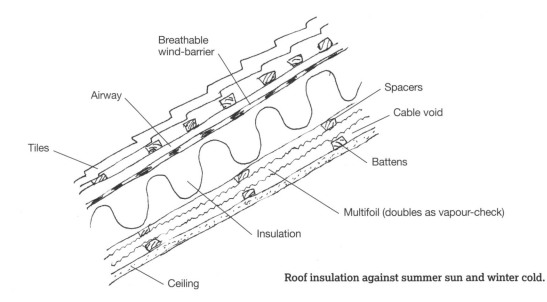

Breathable wind-barrier

Airway

Spacers

Cable void

Tiles

Battens

Multifoil (doubles as vapour-check)

Insulation

Ceiling

Roof insulation against summer sun and winter cold.

able. Also, just as for indoor ventilation, it's important to have many different options for outdoor sitting places: sunny and shady at different times of day, wind-sheltered and breeze-refreshed for different seasons and on paving or grass.

Even in non-extreme climates, we need to both heat our homes and find cool places in and outside them,

according to weather and season. As we know from overseas holidays, warmth and coolness have *qualitative* dimensions: think of enjoying a café frappé in delicious, leaf-dancing shade beside the sea; or lying snug on a Russian stove-bench while snowflakes brush the window. Thermal comfort is a science, but thermal delight is an *art*.

Keypoints

- Bitter winters then stifling summers imply indoor living followed by outdoor living.
- Both heavy insulation with maximised sun exposure and maximum shade with solar heat reflection are needed.
- Deciduous foliage is seasonally matched to sun and shade needs, but opposite to wind-shelter and cooling-breeze requirements. High-crown trees don't impede airflow at person-level.
- Use both the predictability of seasonal winds and adjustable sails and breeze-deflectors.
- Maximise the range of both outdoor sitting-place and indoor ventilation options.
- Rejoice in the qualitative, sensory-aesthetic aspects of warmth and coolth.

Choices

Options	Decision-making factors
Winter windbreaks or summer breezes?	Seasonal wind-directions; breeze-channelling or obstruction; low-level pruning possibilities.
Sealed, open or dual-season house?	Opening-up possibilities; winterising shutters or secondary glazed screens.

Resources

Information

Aronin, J, E. (1973) *Climate and Architecture*, Reinhold, New York.

Cotterell, J. and Dadeby, A. (2012) *PassivHaus Handbook*, Green Books, Cambridge.

Gallo, C. Passive cooling: lessons from the past to present architecture.

Gut, P. and Ackerknecht, D. (1993) *Climate Responsive Building*, BASIN.

Konya, A. (1984) *Design Primer for Hot Climates*, Architectural Press, London.

Kurn, D. M., Bretz, S. E., Akbari, H. and Huang, B., The potential for reducing urban air temperatures and energy consumption through vegetative cooling. Lawrence Berkeley National Laboratory Report LBL-35320, Berkeley, CA

Reed, R. H. (1953) Design for natural ventilation in hot humid weather, Texas Engineering Experiment Station.

Sandifer, S. and Givoni, B. (2000) Thermal effects of vines on wall surfaces: *www.sbse.org/awards/docs/Sandifer.pdf*

US Department of Energy (1988) Landscaping for energy efficient homes.

Willis, S., Fordham, M. and Bordass, B., (1995) Report 31: Avoiding or minimising the use of air-conditioning, BRESCU.

http://eartheasy.com/grow_xeriscape.htm
www.agroforestry.co.uk/forgndg.html
www.coolcalifornia.org/benefits-of-urban-vegetation
www.greendesignetc.net/GreenProducts_08_(pdf)/Stoneman_Josh-Vines(paper).pdf

1. Combination ceiling fan and light fittings can produce stroboscopic illumination: dangerous for the epileptic, and disquieting for all of us. Don't use these.

2. Giovani, B. (2000) Building design for regions with hot climates, in Roaf, S., Sala, M. and Bairstow, A. TIA Conference 2000 proceedings. This system is designed for very hot climates, but also applies to moderate ones.

3. There are several ways to combine these. I describe some in detail in *Spirit & Place* (Architectural Press, 2002).

4. Cotterell, J. and Dadeby, A. (2012) *PassivHaus Handbook*, Green Books, Cambridge.

5. Sandifer, S. and Givoni, B. (2000) Thermal effects of vines on wall surfaces, in Roaf et al. TIA Conference 2000 proceedings.

6. Gut, P. and Ackerknecht, D. (1993) *Climate Responsive Building*, BASIN; US Department of Energy (1988) Landscaping for energy efficient homes.

7. In Maryland, USA: *www.greendesignetc.net/GreenProducts_08_(pdf)/Stoneman_Josh-Vines(paper).pdf* (accessed: 22 May 2014).

8. *www.enviromat.co.uk/green-roof-insulation* (accessed: 22 May 2014).

9. Young, R. (1998) Cool roof, *Building Design & Construction*, February 1998.

10. But not in humid climates with milky night skies. As air is barely cooler at night, it can't cool buildings, so they need low thermal mass. Only lots of air movement (or energy-expensive air-conditioning) cools.

11. Hunter ceiling fans, 2009.

12. There's no definitive thermal tolerance boundary. Besides air movement, humidity and quality, tolerance boundary depends on familiarity, health, age, gender and individual factors. Additionally, the ability to control our environment can increase overheating tolerance by some 6°C (11°F): Willis, S., Fordham, M. and Bordass, B. (1995) Report 31: Avoiding or minimising the use of air-conditioning, BRESCU.

13. Welsh Assembly Government (2009) Climate change strategy – programme of action consultation.

14. And heavyweight ones by 2061: research by ARUP, cited in Masonry construction may benefit from global warming! *Building for a Future*, Autumn 2006, Vol. 16, No 2.

15. So do all wet-trade work as early on in your renovation process as you can.

16. *www.cdc.gov/legionella/about/causes-transmission.html* (accessed: 8 May 2014).

17. Kurn, D. M., Bretz, S. E., Akbari, H. and Huang, B. The potential for reducing urban air temperatures and energy consumption through vegetative cooling. Lawrence Berkeley National Laboratory Report LBL-35320, Berkeley, CA

18. Gallo, C. Passive cooling: lessons from the past to present architecture.

19. Roaf, S. (2001) *Eco-house: A Design Guide*, Architectural Press, Oxford.

20. Tajikistan.

CHAPTER THIRTEEN

Keeping healthy: physical aspects

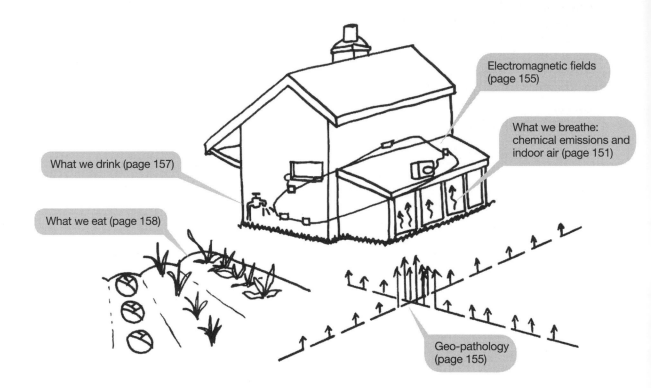

Electromagnetic fields
(page 155)

What we breathe:
chemical emissions and
indoor air (page 151)

What we drink (page 157)

What we eat (page 158)

Geo-pathology
(page 155)

Healthy building: physical/material basics

The environment we're in affects our health. This is so obvious it's a truism. None-theless, outside Scandinavia and German-speaking countries, Sick Building Syndrome (SBS) wasn't recognised until the mid-1980s, and even now its lessons are widely disregarded. SBS is generally associated with workplaces (particularly offices) because sickness-related employee absence costs employers money and workplace-related sickness is fertile ground for litigation. However, it's a serious concern for homes too – perhaps more so, as we spend most of our time in our homes, and during much of this, we're asleep (which is when our bodies replace damaged cells). Also, children spend most of their time at home and, as their organs aren't yet fully developed, they're particularly vulnerable to unhealthy environments.

Buildings affect health through physical, psychological and spiritual means. Physical/material impacts include chemical (mostly gases we breathe), physical (from inhaled fibres and particles to electromagnetism), biological (mostly mould spores and recycled bacteria) and thermal (thermal stress). This is mostly about building materials (see: **What we breathe**, right; **Chapter 6: Insulation materials**, page 59), electric cable layout (see: **Physical but non-sensible factors**, page 155), air-ducting – one reason I prefer natural to forced air ventilation (for other reasons, see: **Chapter 7: PassivHaus,** page 67) – and building condition (see: **Chapter 11: Insulating old walls**, page 119).

There's increasingly a widespread awareness that what we eat affects our health, but we drink more than we eat, and breathe much greater volumes than both of these. The health impacts of what we eat, drink and breathe are, therefore, in inverse proportion to our degree of control over contamination of these nutrients. We can, for instance, grow our own food and filter – although not easily enliven – water, but once a building is completed there's little that we can do to its air quality beyond opening windows, cleaning it, populating it with houseplants and going outdoors. Among building materials, the main chemical offenders are poisons, both deliberate (e.g. wood preservatives, fungicides) and accidental (e.g. formaldehyde, VOCs; see: **What we breathe**, right). For a healthy home, we must therefore add awareness of what we *breathe* to that of what we eat and drink.

Keypoints
....

- Healthy building is even more important for homes than for workplaces.
- This principally involves choice of materials.
- If construction is airtight, this is even more crucial.

What we breathe

We can survive weeks without food, a few days without water, but only minutes without air. In nature, air is refreshed and re-oxygenated by plants (including ocean algae). We call this 'fresh air'. Indoor air, though, isn't freshened; in fact our breathing ages and de-oxygenates it, and 'air fresheners' pollute it.[1] Moreover, it's affected by the materials of which buildings are built, finished, furnished and cleaned. Consequently, wooden houses smell differently to brick houses; natural paints to synthetic paints; leather to faux leather vinyl; and wooden furniture to plastic furniture.

Anything that smells is releasing substance to the air. The quantities off-gassed are usually minute, but that doesn't mean they're harmless. Whereas we only drink two or three litres a day, we breathe some 10,000 litres of air. Its quality, therefore, is important. There are around 55,000 substances used in building materials. Some are harmless and even smell nice (e.g. beeswax, natural – so non-toxic – paint, etc.). Others are known to be harmful, even carcinogenic. We only fully know the health effects of 3% of the things homes are built of,[2] and hardly anything about cocktail combinations, though.[3] If our bodies have always lived with things throughout evolution, it's almost certain they're of safely low toxicity (in normal circumstances). But exposure to unfamiliar chemicals can have unknown effects. That allergies are increasing is widely attributed to environmental pollution (particularly air pollution, most particularly of *indoor* air). It's important not to create a home – however energy efficient – that makes this worse. That's why I always favour natural materials. To some extent, indoor plants can remove airborne chemicals, but it's safer never to introduce toxins.

Timber preservatives are poisons. They deliberately poison insects and fungi, and their liquid, fumes and impregnated dust accidentally poison us. (Also, preservative-treated offcuts mustn't be burnt: arsenic or dioxin smoke isn't good to breathe!) Organic-chemical ones (e.g. Lindane, Dieldrin) are the most hazardous; also copper-chrome-arsenic, although it's more stable. Slightly less toxic but relatively stable are copper-chrome-borax and copper-triazole. It's much safer, however, to use none.

Outdoors, heat-treated softwood resists rot (although it's more prone to splitting, so needs drilling before nailing). So do naturally rot-resistant timbers (e.g. sweet chestnut, oak, European larch, FSC-certified tropical hardwoods[4]). These, however, can be hard to buy in small quantities through builders' merchants.

Indoors, exposed structural timbers not only look attractive but also let you see if they're attacked by insects, so don't need precautionary treatment. If they are attacked, use borate-based preservative. (But, as borate is water-soluble, it's only suitable for indoor use.) As a fire precaution, exposed structural timbers (e.g. posts, beams, joists) need to be intumescent-painted – or to be about 20mm (¾") larger on each exposed side than their structurally required size,[5] so they can char for half an hour before collapsing. Large timbers cost more, but lock up carbon and add both reassuring solidity and sensory-aesthetic charm.

Foremost among accidental chemical hazards is formaldehyde. Despite repeated health warnings over more than 30 years, this is still used in industrial glues, especially in chipboard, medium-density-fibreboard (MDF), plywood, oriented-strand-board (OSB), glue-based carpet and in much furniture. As formaldehyde-based glue is water-soluble, however, *external-grade* plywood and OSB use phenol-based glues, which emit much less formalde-hyde. Try, therefore, to avoid using any glue-based products indoors, or, if you can't, use external-grade ones.

Volatile organic compounds (VOCs) are another major concern, as organic chemicals are highly reactive with our body chemistry. (VOC concentration in homes is typically 10 times that outdoors.[7]) Most VOCs are in the form of solvents (which evaporate – of course!): this is another reason, besides their breatability and non-toxicity, for using water-based, natural-substance-derived paints and cleaning compounds.

Plastics may last 'forever' but aren't wholly stable. PVC, for instance, gives off vinyl chloride and pthalates[8] so, in some German states, is banned in public buildings. Most fitted carpets are plastic and have glued-on backings. They also accumulate dust, dust mites and indoor pollution, and for these reasons have been removed from Swedish public buildings. I consider it wise to avoid such poison-emitting materials and use natural materials wherever possible. This is even more essential in airtight buildings with restricted ventilation: yet another reason for construction that lets the building breathe. It's not the best idea to combine PassivHaus-standard airtightness with PVC windows, fitted carpets, insecticide-treated woodwormy antiques and plastic-foam-padded MDF furniture!

Formaldehyde outgassing (mostly during the first 3-5 years): micrograms per square metre per hour (µg/m²/hr)[6]

Product	Glue	Formaldehyde emission (µg/m²/hr)
Medium-density-fibreboard (MDF)	Urea-formaldehyde	210-2,300
Particleboard (chipboard)		100-2,000
Furniture-grade / interior-grade plywood		7-1700
Furniture-grade plywood with vinyl or laminate surface		3-300
Construction-grade plywood, external-grade	Phenol-formaldehyde	2-83
Oriented-strand-board (OSB), external-grade		

Healthy and unhealthy materials: general principles

Healthy materials	Unhealthy materials
Wood	Glued products (e.g. carpet backing, chipboard, MDF)
(Real) linoleum	Vinyl
Clay products	Plastics
Natural fibres	Mineral fibres
Natural paints, water-based	Synthetic and VOC-based paints
Lime (for indoor climate; but it's a potentially hazardous material to work with)	Cement (for indoor climate)

The physical characteristics of things we breathe also affect us. The most notorious of these are dust and dust mites for asthma and asbestos for asbestosis. Glass fibres are sharp, although not as sharp or as impervious to blunting as asbestos; and can break into short, hence breathable, lengths. I therefore avoid glass fibre insulation and reinforcement (e.g. in plasterboard).

At a molecular scale, fresh air is rich in negative ions. These clean air negatively ionised air molecules lack electrons so are 'hungry': they attach themselves to micro-dust and settle it to the ground. Negative ions also affect our well-being.[9] Their scarcity in pre-thunderstorm air causes headaches and irritability. Pollution, magnetism (e.g. from steel air-ducts), electromagnetism and friction (e.g. from forced-air fan motors) destroy negative ions so air is less clean and we feel less than well. This is one reason I prefer natural ventilation; and never use mechanical ventilation for inlet air.

If you move into a new building impregnated with toxic chemicals, the only things you can do to detoxify it are fill it with houseplants or keep windows open – and lose lots of heat. But if you build a new eco-house, you can avoid almost all of these toxin sources. (You can even use LSF (low-smoke-and-fume) type electrical cable instead of PVC.[10]) Indeed, but for some plastic appliances, you can avoid all such toxic gas-emitting materials.

Choices

Options	Decision-making factors
Materials: synthetic or natural?	Price and ease of maintenance versus occupant health and environmental impact.
Preservative-impregnated timber or rot-resistant design?	Availability of durable woods; availability of low-toxin preservatives; contractor's understanding of rot-resistant construction.
Natural or artificial ventilation?	Indoor air quality; heat loss/recovery; control; fan noise; outdoor noise.

Keypoints

- Avoid plastics (especially PVC) and toxic-gas-emitting materials wherever possible; also glass fibres and fitted carpet.
- Avoid formaldehyde-based glued products (especially chipboard, MDF and internal-grade plywood).
- Use natural materials wherever possible.
- Use heat-treated or naturally rot-resistant (FSC-certified) timbers instead of timber preservatives.
- Use natural ventilation, and preferably breathing construction.
- Use houseplants to help clean air.

Indoor plants to detoxify air[11] (darker shading = remove more toxins)

Plant species	Chemical absorbed		
	Formaldehyde (from combustion (e.g. tobacco smoke), plywood, chipboard, MDF, glued materials, cleaning materials)	Benzene (from combustion (e.g. tobacco), plywood, chipboard, adhesives, mastic, cosmetics, deodorisers)	Trichloroethylene (from paints, varnish, adhesive, mastic, cleaners, correction fluid)
Aglaonema Silver Queen	▨	▨	
Azalea	▨		
Evergreen Palm, *Chamaedorea seifrizii*	▨	▨	▨
Chrysanthemum, *Morifolium*	▨	▨	▨
Dieffenbachia	▨		
Dragontrees, *Dracaena Deremensis Warnerkei*	▨	▨	▨
Dracaena marginata	▨	▨	▨
Dracaena Massangeana	▨	▨	▨
Dracaena Janet Craig	▨	▨	▨
Weeping Fig, *Ficus benjamina*	▨	▨	
Perennial Barberton Daisy, *Gerbera Jamesonii*	▨	▨	▨
Goldheart Ivy, *Hedera helix*	▨	▨	▨
Elephant's Ears: *Philodendron domesticum*	▨		
Philodendron oxycardium	▨		
Philodendron selleum	▨		
Mother-in-law's tongue, *Sansevieria Laurentii*	▨	▨	▨
Scindapsus aureus	▨	▨	▨
Peace Lily, *Spathiphyllum*	▨	▨	▨

Physical but non-sensible factors

There are things we can't sense that affect health, including radioactivity and electromagnetism. Protection measures against ground-source radon are now routine. Some building materials from deep-earth origin contain uranium, which is incombustible, and therefore concentrated by burning. Uranium releases radon (although much less than some ground). As a result, blast-furnace slag or pulverised fuel ash insulation blocks can be twenty times as radioactive as bricks;[12] and plasterboard of phosphogypsum (a by-product of fertiliser production), a hundred times more radioactive than gypsum plasterboard.[13] In airtight buildings, this matters. Recycled waste is attractive ecologically, but not always healthy to live with.

There are also concerns about electricity, electromagnetic fields (EMF) and geo-pathic radiations[†]. To avoid geo-pathic locations, avoid positions where trees show abnormal distortions. Trees often lean away from strong electromagnetism (e.g. transformer stations). Walls and floors obstruct electric fields[†], but don't reduce electromagnetic fields. Consequently, it's important to keep the whole house – especially bedrooms – as far away as possible from transformers, power lines and telecommunication masts.

Electricity is everywhere in modern life. Many people find they sleep better away from electricity. Some are more sensitive than others. Basic precautions include distancing electricity from beds by 1.2m (4') (e.g. by controlling bedside lights with pull-switches) and even further from consumer units and (old-style) televisions. Ring mains reduce resistance to electricity flow – so reduce energy consumption, copper use and fire risk – but current is drawn through the whole ring whenever any appliance is on, so generates EMF all round the house. Spur layouts limit most EMF to single spurs but may require heavier grade cable for fire safety. Cable insulation obstructs, but *doesn't eliminate*, electricity flow to earth, so some EMF is always produced. Much cable is routed through ground-floor ceilings, namely bedroom floors. To counter this, you can use 'demand switches' which turn off *all* electricity when appliances are off. Auton-

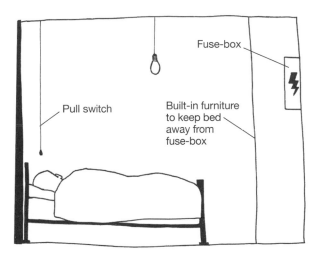

Distancing electricity from beds.

omous appliances (e.g. refrigerators, central heating) need separate circuits so they don't make all wiring live when they switch themselves on.

All this, of course, is disputed by vested interests: manufacturers, power and telecom providers and the air-conditioning industry. It's possible they're right,[14] but I follow the precautionary principle.

Keypoints

- Avoid recycled furnace/ash materials as these increase radioactivity.
- Distance all electricity from beds by at least 1.2m (4').
- Avoid placing your house where trees show abnormal distortions.

Choices

Options	Decision-making factors
Recycled (but radioactive) materials or virgin materials?	Environmental impact versus occupant health; ventilation rate; amount; source and manufacturing process.
Ring mains, spur layouts and/or 'demand switches'?	Electro-sensitivity; shallow sleep; cable cost; capacity for subsequent extensions and/or additional sockets.

Making an old building healthy

When eco-renovating an old building you'll need to bear in mind all the issues discussed above, but the main health difference between old and modern buildings is their chemical content. In addition, however, older buildings have a few specific considerations of their own.

Traditional buildings were generally built of natural materials and, as a general rule, have fewer health-endangering chemicals in their building fabric. However, some less old buildings were built using some materials so hazardous that they're forbidden now – these you need to be aware of and remove wherever possible. Lead, a brain poison, is one. It was commonly used in old water pipes. Also, until the 1960s, white oil-paint (common on wooden external walls and window-frames) contained lead:[15] this is particularly dangerous for children as dust from peeling paint can get on to their hands, so into their mouths. Consequently, treat peeling old paint as toxic. Until the 1980s, the wonder material, asbestos, was widely used but is now known to be highly carcinogenic. Its manufacturers claimed it was safe (and have subsequently been bankrupted by compensation claims). Asbestos-cement drainpipes were common in the 1950s, and asbestos-cement 'slates' until the late 1980s. Also, pre-1990 Artex (decorative plaster) is likely to contain asbestos.

Few synthetic chemically based materials were used prior to the 1950s, since when their number and amount have grown exponentially. Old buildings, however, may have suffered 'improvements' by subsequent owners, which introduced harmful substances (e.g. timber preservatives, mercury-based anti-fungal paints and wallpaper pastes and chipboard-cored kitchen cupboards). Additionally, many have stood long enough to have been attacked by rats. And prolonged dampness may have caused mould, the spores of which are seriously unhealthy. Many of such health failings, however, are easily remedied.

Rats can chew through almost anything even slightly soft. They usually, however, enlarge existing apertures (e.g. where drains pass through walls). This is a location for hard cement mortar (separated 2-3mm from pipes so walls and pipes can move independently from one another). For rat holes in lime mortar, push in broken glass then re-point. Ultrasound successfully deters rats. The best deterrent is a cat or terrier (or fox urine – but this is also a human deterrent).

Mould can be cleaned off with white vinegar or a vinegar and borate solution.[16] To prevent its recurrence, however, you need plenty of fresh air to dry everything out. Sunlight is also fungicidal.

Replace lead water piping – you don't want to go the way of the ancient Romans (once they started to drink from lead-lined water channels, bye-bye Roman Empire).[17] Lead paint should be carefully scraped off (if outdoors, over a dust-sheet), vacuumed up, bagged and disposed of as toxic waste. It's safest to leave asbestos untouched, but Artex can be overcoated with filler or steamed then scraped and bagged: allegedly safely,[18] but I'm not convinced. (Wear a dust-mask and overalls you can wash separately from other laundry!) Old chipboard has probably off-gassed most – but not all – of its formaldehyde: which has now impregnated all soft furnishings. I recommend removing both it and them.

Despite such problems, old houses have fewer unstable chemicals in them than most modern ones, so they're generally healthier to live in. Nonetheless, indoor air quality must be a prime concern in all types and ages of buildings. It's easy to make indoor air quality worse, so be very careful in choosing what materials you use to improve the appearance and energy performance of old buildings.

Keypoints

- Older buildings were plumbed with lead water pipes; and external wood was often painted with lead paint.
- Until the 1980s, asbestos was widely used.
- Mould and rodents are common. Both can be remedied.
- Otherwise, except for subsequently applied preservatives and products, old buildings contain little toxic material, so are generally healthier than modern ones.

Choices

Options	Decision-making factors
Remove, cover up or leave asbestos undisturbed?	Risk of airborne fibres; stability if undisturbed; expense of specialist removal.
Remove or leave old chipboard?	Age; humid environment; ventilation adequacy; sealing/covering possibilities; opportunity to use natural wood replacement.

What we drink

Our bodies are about 70% water, so the quality of the water we drink directly affects our health. Mains water is filtered and treated (usually with chlorine: poisonous to both microorganisms and us). However, as water is 'a universal solvent', we wash things we don't want – both biological and chemical – into it. Moreover, much urban mains water has already been drunk or used before: it's recycled (a euphemism for shat into). Consequently, the 1988 Nader report on US drinking water found 2,000 chemical contaminants, of which 200 are carcinogens, mutagens or similar.[19] Although biological pollutants are dealt with by sewage treatment, chemical, heavy metal and pharmaceutical pollutants tend to remain.

If you're worried about this (in many places, I'm not), these pollutants can be removed at home by activated carbon or ceramic filters or by reverse osmosis. As these systems reduce water pressure and cartridges have a limited life, they're usually only installed before the kitchen tap, so are housed under the sink. Bottled water is pollutant-free but, as water's life-energising capacity declines in storage,[20] is of arguable benefit – not to mention its environmental costs from packaging and transportation.

'Hard' water (e.g. from limestone areas) is fine to drink but deposits limescale, which builds up over time. If this clogs pipes or damages appliances, it can be remedied by fitting an in-line water softener (e.g. a container filled with anionic resin granules) on the way to the cold tank (which supplies all water except the kitchen cold tap, which we drink from). If you haven't a cold tank,[21] place the in-line water softener as near to the supply entry point as practical. Strongly acid water (e.g. from peat bogs) can dissolve copper pipes, leaving blue-green toxic copper oxide in the water. To protect *pipes*, you need an in-line de-acidifier (e.g. a container filled with magnesium oxide granules). To protect *yourself*, it's advisable to supply the kitchen cold tap only through stainless steel or MDPE piping (e.g. alkathene).[22] Some artesian borehole water contains heavy metals, including arsenic, so needs to be tested for chemical, as well as biological, contamination.

Keypoints

- Protect pipes, fittings and appliances from strongly acidic or alkaline water.
- If water is significantly acidic, don't use copper piping to the kitchen tap.
- If worried about water contamination, fit a filter.
- Test private water supply for biological pollution; and artesian water, also for heavy metal pollution.

Choices

Options	Decision-making factors
Filter water?	Source; recycled content; industrial/sewage/fracking pollution potential; chlorination/fluorination; contamination by old supply pipes; water pressure.
Acidity/alkalinity correction	pH; susceptibility of pipes/fittings/appliances to damage; water pressure.

What we eat

Our bodies are built of what we eat. Much of the food we can buy is grown with artificial fertilisers, so is nutritionally unbalanced, may contain biocide residues and is of suspect nutritional value. Some even contains (undeclared but legally permitted) genetically modified material. Food accounts for 17% of the CO_2 human activities cause, part of which is due to energy-intensive artificial fertilisers, part due to wastage, and part to the huge distances it travels: in Britain, food transport alone produces 19 million tons of CO_2 per year.[23] From a health perspective, if food can't grow here, it can't be fresh or naturally ripened; is sometimes coated, treated or irradiated[24] to survive the journey, and isn't traceable or subject to national quality controls. It's questionable, therefore, how much good it does us.[25] Such realisations have led to a resurgence of organic food-growing at home. However, there's also a more ominous issue: even in the near future, *will there be enough*? Climate change and oil depletion will reduce agricultural yields, at least using current agricultural systems, and many fear that this may lead to wars. Growing food locally, even in allotments or private gardens or even on balconies, roofs, in 'window-greenhouses' and (for fresh greens) window boxes, could become an important part of future food security.

How easy is it to grow food at home? Meat isn't easy: livestock require constant attention (e.g. chickens must be shut in *every* night; miss one night and they're fox-food) and many towns have bylaws prohibiting livestock. Growing vegetables, however, *is* easy. They need somewhere to grow (preferably in soil, as they naturally would), nutrient, light, warmth, water and occasional attention (e.g. weeding, protection from pests).

Few house plots have nutrient-rich soil, but this is easily improved by compost (see: **Chapter 17: Things we get rid of**, page 197). Light largely depends on orientation and plot design: unshaded south-facing slopes get most. South- and west-facing walls (including house walls), being most warmed by the sun, suit grape vines and cordon fruit. If you get frosts in springtime, west-facing walls are, in fact, better for fruit growing than south facing. (The hardest frost is at dawn – just before the sun appears: on east-facing walls, the rising sun would kill frozen blossoms but on west-facing walls they have all morning to defrost before being touched by sunlight.) Warmth is closely

Window-greenhouse.

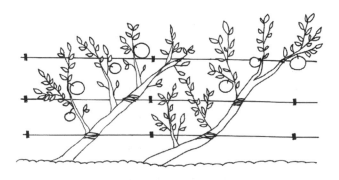

South- and west-facing walls suit cordon fruit growing.

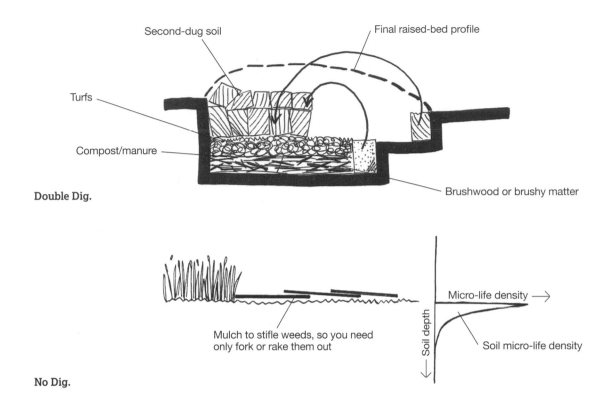

Double Dig.

No Dig.

related to sunlight exposure, but also involves wind-shelter. Also, it's not a good idea to lay heat-pump heat-collection coils under a vegetable bed, as these cool the soil. For water, rainwater is best – and free. It only needs simple rainwater butts (e.g. recycled plastic drums). Roofs, however, mustn't be made of toxic or phyto-toxic[†] materials (e.g. lead, copper or extensive bitumen) as this will poison the water, then the soil, the crops, then us.[26]

For starting a garden, there are two contrasting schools of thought: Double Dig or No Dig. Double Dig optimises conditions for root growth (and buries weeds so deep they won't sprout). No Dig minimises damage to soil micro-flora, each kind of which thrives best at a particular depth. It's also *much* less hard work (although you may have to weed more later). The easiest way I've found to prepare new ground for cultivation is to mulch it (e.g. with packaging cardboard) until all the weeds have died, then fork them in or rake them off to compost.

Greenhouses and polytunnels extend growing season and hugely boost yields. Used for starting seedlings, they help outdoor crops get ahead; and their hot-house climate broadens the range of species it's possible to grow. Polytunnels are inexpensive but depend

Indoor or outdoor growing?

Outdoors	Greenhouses/polytunnels
No embodied-energy / pollution whatsoever.	Embodied-energy / pollution
Need infrequent attention (slug-hunting excepted).	Need frequent attention (water, temperature, greenfly, etc.).
Suit staples, particularly root-crops.	Suit sensitive/exotic vegetables; start seedlings early; extend growing season.

Greenhouse/polytunnel or sun-room / conservatory?

Greenhouse/polytunnel	Extended growing season and produce range.	North–south axis
Sun-room/conservatory	Starting seedlings for earlier crops.	South facing

on short-life, disposable polythene: not very ecological! Greenhouse glass is fragile and needs protection from strong winds and playing children. Although chicken-wire screens can protect from balls and, to some extent, from wind, this limits siting options. Sun rooms and conservatories can double as greenhouses, although for greenhouses, freestanding north–south axis maximises summer sunlight, so is better. Greenhouse plants can increase shade and humidity indoors and bring occasional plagues of aphids and suchlike. These, however, are summer problems so can be resolved by opening windows and accepting shade. Openable glass screens between conservatory and living room also help.

We're not the only ones who eat garden produce; many pests do. Rabbits and racoons, for example, can eat huge amounts in a single night. As these animals can dig under normal fences, the wire-netting needs to continue underground (6"/150mm down, 12"/300mm out). As racoons can also climb fences, these need floppy chicken-wire tops, overhanging outwards – as for keeping cats from children's sandpits.[27] Slugs and snails, however, are often the worst pests. Slug pellets are coloured to discourage animals from eating them, but they do, and birds eat poisoned slugs. However teeth-grindingly annoying are crop-annihilations by slugs, it's always worth remembering the harm such poisons do. Nor are they necessary. As slugs need damp soil to slither over, watering in early mornings instead of at night can greatly reduce their numbers. Dawn and dusk slug patrols can catch bucketfuls, especially if you've placed 'delicious' old lettuce leaves for them to hide under. Copper strips around individual plants, lime circles on dry soil and salt barriers on concrete paths are also effective. (Salt isn't a good idea on soil: it washes in, so makes soil too saline. Similarly, don't over-use lime.) Then there are 'organic' slugicides, non-toxic to birds and animals. The most effective slug protection that I have found is to combine Swiss slug fences (which slugs can't climb over) with nematodes (which invade slugs' bodies and eat them from the inside). Besides all this, a simple pile of brushwood can attract hedgehogs – which can eat half their body-weight in slugs per day.

If you grow food, you'll need to store it. This affects house design. Few freezers are big enough, and anyway they don't suit frost-sensitive produce (e.g. potatoes). Moreover, refrigeration takes 15% of household electricity.[28] And the combination of rising petroleum price and increasingly severe weather

To protect food crops from rabbits and racoons, wire-netting fences need to continue underground.

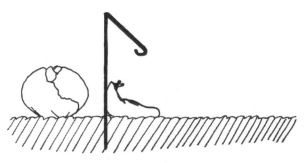

Swiss slug fences: slugs fall off when trying to slither across the fold.

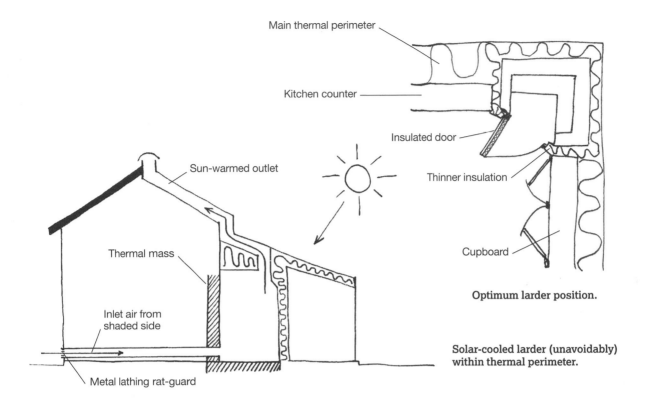

Optimum larder position.

Solar-cooled larder (unavoidably) within thermal perimeter.

extremes make power-cuts increasingly likely.[29] Houses need larders: well ventilated and preferably on north walls (for shade) and outside the thermal perimeter. External stores (sunk into the ground for frost-free coolness) or cellars are other possibilities.

Health involves more than just what we eat, drink and breathe; indeed, it depends on a healthy spirit. Nonetheless, healthy food, water and air unequivocally provide its material foundation. For a healthy life, therefore, these are essential prerequisites.

Keypoints

- Non-toxic roofs can feed rainwater butts for irrigation water.
- South- and west-facing walls suit cordon fruit growing. West walls best protect against late frosts.
- Sun-rooms can double as greenhouses.
- Brushwood piles attract slug-eating hedgehogs.
- Copper strips and lime circles around plants deter slugs; Swiss slug fences keep out invading slugs; nematodes kill them.
- Larders should be cool, so are best on north-east corners.

Choices

Options	Decision-making factors
Indoor or outdoor growing?	Staple or exotic vegetables; length of growing season; climate; attention.
Greenhouses or polytunnels?	Longevity; fragility; environmental impact; price.
Sun-rooms as greenhouses?	Aphids, etc.; summer shade indoors; humidity indoors.

Resources

Physical aspects: general

Bachler, K. (1989) *Earth Radiation*, Wordmasters, Manchester.
Bower, J. *The Healthy House* (4th ed.) The Healthy House Institute.
Curwell, S., March, C. and Venables, R. (1990) *Buildings and Health: The Rosehaugh Guide*, RIBA Publications, London.
Hall, A. (1997) *Water, Electricity and Health*, Hawthorn Press, Stroud.
Hunter, L. Mason (1990) *The Healthy Home*, Pocket Books, New York.
König, H. (1989) *Wege zum Gesunden Bauen*, Ökobuch, Freiburg.
Stakeholder Advisory Group on ELF/EMFs: UK Department of Health.
www.dirtyelectricity.ca/electrical_health_implications.htm

Less common products

Demand switches: *www.equilibrauk.com/acatalog/Demand_Switches.html*

Air

Non-toxic indoor paints
Aglaia: www.naturalpaintsonline.co.uk
Auro: *www.organicnaturalpaint.co.uk/auro*
Livos: *www.livos.co.uk/*

Non-toxic outdoor paints
Keim mineral paints: *www.keimpaints.co.uk*
Osmo wood care: *www.wood-finishes-direct.com/Osmo*

Food growing

Morrow, J. (2014) *Vegetable Gardening for Organic and Biodynamic Growers*, Lindisfarne Books, Great Barrington, MA.
Seymore, J. (2nd ed. 2008) *The New Self-sufficient Gardener: The Complete Illustrated Guide to Planning, Growing, Storing and Preserving Your Own Garden Produce*, Dorling Kindersley, London.
www.biodynamic.org.uk
www.permaculture.org.uk
www.soilassociation.org

Pests

http://ucipm.ucdavis.edu/PMG/PESTNOTES/pn74116.html
www.gardendesk.com/2009/04/3-ways-to-keep-raccoons-out-of-corn.html
www.rhs.org.uk/advice/profile?pid=228
http://eartheasy.com/grow_nat_slug_cntrl.htm

Water

www.cdc.gov/legionella/about/causes-transmission.html
www.ciphe.org.uk/Global/Databyte/Safe%20Hot%20Water.pdf

1. Indeed, air-freshener spray cans carry health warnings. Nor are they necessary: to clear toilet smells, just strike a match to ionise the air.
2. Liddle, H. (2010) Have we become too fixated on modern materials?, *Building Design*, 5 February 2010.
3. Under the US Toxic Substances Control Act (1976), 62,000 chemicals were 'grandfathered' (i.e. permitted due to common use). Around 20,000 more have since been added. Of these, 85% have no health data; 67%, no data at all: *www.greensciencepolicy.org/node/26*
4. FSC certification, however, excludes small producers (who can't afford certification costs) and pine-beetle-killed timber (whose rotting would produce CO_2).
5. *www.mace.manchester.ac.uk/project/research/structures/strucfire/materialInFire/Timber/Charring/standardFires.htm* (accessed: 5 September 2014).
6. Bower, J. *The Healthy House* (4th ed.) The Healthy House Institute.
7. de Selincourt, K. (2014) Healthy buildings must be warm, well ventilated and dry, *Green Building* magazine, Spring 2014, Vol. 23, No. 4.
8. Curwell, S., March, C. and Venables, R. (1990) *Buildings and Health: The Rosehaugh Guide*, RIBA Publications, London.
9. Article from *Journal of Aviation, Space, and Environmental Medicine* (August 1982, pages 822-3).
10. *www.greenbuildingforum.co.uk/newforum/comments.php?DiscussionID=4208*
11. *Perspective* Sept/Oct 1993; Curwell et al. *Buildings and Health*; Planverkets Rapport 77 (1987) Sunda och Sjuka Hus, Statens Planverket, Stockholm.
12. German figures from König, H. (1989) *Wege zum Gesunden Bauen*, Ökobuch, Freiburg.
13. Curwell et al. *Buildings and Health*. Fortunately, most plasterboard is thin: a normal partition wall contains 25mm (1") of solid matter, a quarter of that in a 110mm (4 ½") wall.
14. This I must say to avoid being sued! It's not, however,

obligatory to believe it.

15. Lead was used in other colours too. Its use in paint wasn't actually banned until 1997 (USA), 1992 (UK) and 2005 (Canada).

16. Hunter, L. Mason (1990) *The Healthy Home*, Pocket Books, New York.

17. The fall of the Roman Empire has been variously attributed to a number of causes. Lead poisoning is but one of these – but it is chronologically consistent.

18. *uk.answers.yahoo.com/question/index?qid* (accessed: 7 December 2013).

19. Hunter, *The Healthy Home*.

20. Wilkes, J. (2011) Art in the service of nature, *Art Section Newsletter*, 2001, Issue 35; Eble, J., unpublished lecture, at Living Architecture Conference 1981, Järna, Sweden.

21. Some houses have no cold water tank, so use mains-pressure water supply for everything.

22. Be aware that stainless steel comes in many grades. Some aren't stainless, some can leach nickel: *http://nickelfreelife.com/nickel-allergy-faq.php* (accessed 23 November 2014).

23. *www.climatechoices.org.uk/pages/food3.htm* (accessed: 9 December 2013).

24. *www.food.gov.uk/science/irradfoodqa/* (accessed: 23 May 2014).

25. Also, some people think that we can't be fully nourished by food that hasn't grown in the same cosmological environment that we live in.

26. Wack, H.-O. Rainwater utilization in housing, Wisy.

27. *http://ucipm.ucdavis.edu/PMG/PESTNOTES/pn74116.html*; *www.gardendesk.com/2009/04/3-ways-to-keep-raccoons-out-of-corn.html* (accessed: 7 May 2014).

28. Energy Saving Trust, Powering the nation report CO332(1).pdf (accessed: 11 January 2014).

29. Once frozen then thawed, food doesn't keep long. Ice crystals have stabbed it through, opening up a multitude of routes for invading bacteria.

CHAPTER FOURTEEN

Keeping healthy: spirit and soul

Nature connection
(page 176)

Daylight and mood (page 165)

Sensory nutrition
(page 170)
Soul-nourishment
(page 173)

Space quality
(page 167)

Healthy building: supra-physical basics

Human health and sickness isn't just about germs. Non-infectious illnesses also cause deaths and infirmities: in some countries more than do bugs.[1] Anyway, sickness is a far more complicated issue than simple material cause and effect. It is, after all, about *life* – which isn't material. Material bodies are only vehicles for this. In fact, we're multi-layer beings. We have physical bodies that respond to physical abuses (including, but not only, pathogens). We're alive – so subject to a whole range of nutritional influences, not only material. We each have a soul – which is nourished or poisoned by place-mood and sensory, social and emotional experiences; and a spirit, which recognises embodied spirit – positive or negative – and its manifestation in beauty or ugliness (which transforms or crushes us inwardly).

Physically, sharp hard shapes are painful to bump into; hard floors jar the spine; and design that requires us to make awkward movements exhausts us or strains muscles. At a life level, our bodies are electro-chemical systems, so electricity,[2] electromagnetism,[3] radioactivity and (some experts say, and apparently our ancestors believed[4]) cosmic and geopathic radiations[5] affect us. This is in addition to physical and thermal nutrition – or poisoning (see: **Chapter 13. Keeping healthy: physical aspects**, p.150). Soul-nourishment involves sensory experiences and their resonations in the soul. Spirit-nourishment includes embodied values, the appropriateness of what places and the materials that they're made of 'say'; aesthetics at a universal, not stylistic level; and beauty – which helps us transcend material concerns. We also respond strongly to archetypal memories: security, ensocialising fire and, especially, nature. Indeed, for the spirit, nature is a great, perhaps the greatest, healer, particularly in times of stress.

Daylight and mood

Just as plants need sunlight, so do we. It nourishes both soul and body. Sunlight activates photo-chemical processes, upon which life depends. This has an aging effect on timber, fabrics and skin, and can trigger out-of-control cell division (particularly melanoma) and accelerate other cancers. Importantly, though, it's bactericidal and vital for vitamin D production, liver processes and hormone-regulating organs (pituitary, pineal and hypothalamus) affecting melatonin, growth, balance – and it's vital to lift mood.[7] Our spirits are greatly affected by the quality, strength and duration of daylight, sunlight's spectrum-related daughter. Seasonal Affective Disorder (SAD) has both physiological causes (e.g. hormones affecting circadian rhythms) and shorter-term mental causes (e.g. gloom makes us feel gloomy). Consequently, sunlight 'lightens' mood, even in hot climates; that's why so many people holiday in them. Besides providing zero-energy lighting, daylight is a soul-nourishing asset. Daylight also benefits rooms (e.g. bathrooms, passages, staircases, draught-lobbies) that don't need views. As

Keypoints

- Human beings are multi-layer beings: physical body, life energy, soul and spirit.
- All this affects health: ergonomics and physical impacts; food, water and air; place-mood and sensory nutrition; and beauty (or ugliness) and embodied spirit.
- Our living bodies are electro-chemical systems, so electricity, electromagnetism and ionising radiation affect us.
- Nature heals the spirit.

Who are we? Our multiple layers of being			Environmental qualities needed for health
Spirit	We *recognise* it, but can't define it.	Invisible and intangible: but the core of 'us'.	Embodied values, transformative beauty.
Soul	We *feel* it, but can't define it.	Invisible and intangible: but without feeling, living isn't living.	Sensory and social/emotional attractiveness, embodied feeling/soul.
Life	We *know* what it is, but it eludes definition; e.g. when does life start and stop?[6]	Invisible (although sensible): but without this bodies are useless.	Healthy air, water, food, warmth and light. Energising (invigorating and/or recuperative) surroundings.
Body	We know what it is, and can *define* it.	Visible and tangible: but not 'us'. (Amputations don't diminish our identity.)	Ergonomic, functional practicality.

Upstairs, light-tubes to rooms below can be 'boxed in' and hidden in corners, or within built-in cupboards.

outer walls are sometimes too valuable for views from principal rooms, roof-lights, light-tubes or over-door windows can be used to light these non-view rooms.

Normally, however, daylight costs heat loss. Even triple-glazed windows lose heat. This (and expense) limits window area. Consequently, daylight indoors is rarely more than a tenth of that outdoors. In some circumstances, this reinforces cosiness and security. (Small children, in particular, need twilit spaces to 'nest' in and free their imagination from the over-materially informative brightly lit world.) In other circumstances, though, gloom makes us feel trapped. Moreover, on dull days, a tenth of outdoor light doesn't feel enough – nor is it enough to counter SAD. (Although it's recommended to spend at least an hour in outdoor daylight every day, few of us can.) Splayed window reveals, however, admit more light for no more heat loss – and, by introducing a mid-tone between bright window and shaded wall, also reduce glare. Above and beside windows, there should be enough wall-space so blinds, curtains, pelmets, etc. don't encroach on window area. Coloured surfaces, especially under windows, colour the light they reflect. Generally, warm hues bring soul-warmth to a room's mood. Pale tones brighten spaces making them feel bigger. Dark tones do the reverse: light and mood

both becoming gloomy. (Dark-toned kitchen counter-tops are a common offender here.)

Windows in more than one wall distribute light better and lighten shadows. Moreover, the interaction of ever-changing colours and strengths from different orientations and times of day and year gives much more 'life' to light than can mono-orientation windows. Wherever possible, therefore, rooms we spend most time in should have windows in two (or more) walls. This combination of multi-aspect, colour reflection and tonal graduation raises daylight from the merely visual-informative to the soul-nourishing.

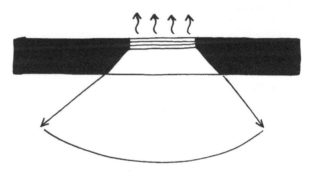

Splayed reveals admit more light without increasing heat loss.

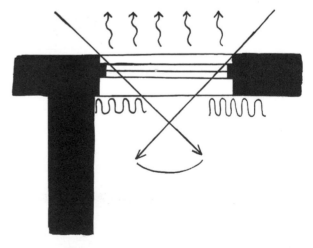

Windows obstructed by curtains admit less light but still cool as much.

Choices

Options	Decision-making factors
Mono-aspect or dual-aspect fenestration?	Energy conservation; building form; light quality; proximity of boundary.
Interior windowsill colour?	Mood of reflected light; material; cleanability.
Service-room(s) location?	Need for view; non-window daylighting possibilities; orientation/warmth of light from sky.

Spatial factors

Space isn't the same as spaciousness. Nor is it about number of rooms. Indeed, number of rooms is more about how we think about space than about what space does to us. There are two aspects to space: physical and psychological. *Physically*, too little space is irritatingly cramped; too much, frustratingly inefficient. Besides such impracticalities, inappropriate *psychological* space prevents us feeling at ease. All these factors contribute to stress, which can lead to physical illness or even divorce.

Physical space affects the efficiency of daily life functions. There are three kinds of physical space: space for doing things in, space for moving in and space for keeping things in. Space is expensive to build and, worse, *to heat*, so it makes sense to reduce physical space to the minimum required for efficiency. For storage, we should distinguish between things we use hourly (which need shelves), daily or weekly (cupboards) and seasonally – which are cheaper to store in bays in a communal shed than at home.

Not enough space to do things in means we only half live. Not enough space to put things in means we feel trapped by clutter. Too much space dilutes social

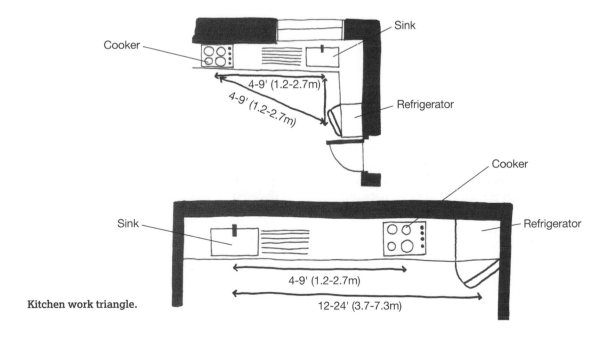

Kitchen work triangle.

Light and spaciousness is relaxing.

A home's social heart needs compressed, focally warmed space.

cohesion. Poorly arranged space to move in means we waste time and effort, and bump into each other, promoting social friction.

Kitchens are frequent offenders here. Most movement is between three key points: the sink, refrigerator and cooker. Sinks are usually best positioned by windows to allow a view out while washing up; refrigerators need to be where they aren't warmed by sunlight or stoves; and cooker hoods need ventilation outlets, so cookers are best placed by exterior walls (or approximately under the roof ridge). Despite such limitations, these three key points should be only a few paces apart (4–9'/1.2–2.7m); and the total triangle, 12–24' (2.6–7.2m) without obstructions.[8] They can be arranged in one straight line, parallel (with cupboards and refrigerator 4'/1.2m across from sink and cooker) or, best, in an L- or U-shape or curve.

Psychological space is different from physical space. It affects both soul and spirit. We need spaciousness to feel free in, let the mind expand and the spirit soar. Uncluttered spaciousness is about freedom from pressure. It's calming; it relieves stress. However, this isn't only about generous space to move in, so doesn't necessarily require expensive floor space. Low windowsills let the soul breath out. (For safety, however,

low-silled upstairs windows need child-proof opening restrictors, and any glass below 32"/800mm above the floor should be toughened.[9]) Height that we can't use also makes rooms feel spacious and lifts our spirits. Nor indeed is it solely about dimensions: lots of light indoors, visually calming simplicity and auditory and textural quietness have a similar effect.

Space-need, however, is also much influenced by context, and is often reactive (e.g. wilderness cabins

Where do we need spaciousness? Where, cosy compact spaces?

Cosy	Spacious
Socially cohesive spaces, e.g. hearth, dining.	Spaces for physical activity, e.g. living room, kitchen, workshop, games room.
Secure spaces to retire to, e.g. bed, sofa, armchair.	Mental relaxation and stress-relief recuperation spaces, e.g. living room, balcony.
Protective withdrawal, e.g. personal alcove, toilet (to sulk in), children's nooks.	Waking up to face a new day with optimism, e.g. bedroom, breakfast area/ balcony.

typically need to feel cosy and protective; urban apartments, spacious and calm: the exact opposite of land cost). Context also affects the entry experience: to leave the outer world behind, is a front door opening straight into the living room sufficient? Or do you need a gate, front garden, porch and hallway? These are things for the whole family to discuss.

Price often restricts spaciousness to where it matters most. Even if affordable, however, not *all* rooms should be generously large. Small children enjoy small spaces (from nooks and open-fronted cupboards to under-stairs) to 'nest' in. Also, excessive space can make homes feel un-lived-in, socially cold. Sometimes we need compaction to feel snug, cosy and secure. For fireside cosiness, we need to feel thermally, aromatically, audibly and visually warmed by flames: important elements of the 'hearth' experience.[10] Cosiness also implies placing the stove in a darker corner or alcove, small and low enough to compress *social* warmth. (In a barn-sized or daylight-flooded room, even a blazing fire can't create a cosy mood.) Traditional homes combined the 'mothering' warmth and baking aromas of cooking ranges with this social heart. Kitchens we eat in can still do this, if made of sensorially inviting, not sterile 'hygienic', materials. For both soul-mood and energy conservation, warmth – and its supportive sensory

and spatial character – should centre the house. Indeed, virtually all soul-nourishing features also make ecological sense. Nestling into the landscape, for instance, sheds wind; and compact houses both feel secure and snug and minimise heat loss (see: **Chapter 4. Choosing a site**, page 31).

The key issue is the *mood* an activity – hence room – needs. Which activities need spaciousness, which cosiness?

We often do different things in one place at different times (e.g. cooking, eating and homework in the kitchen). Sometimes this is the only, and unsatisfactory, option, but more often these activities share a particular mood. Some functions, however, need specific moods. These need to be distinct parts of rooms (e.g. delineated by beams, floor material, light), alcoves or separate rooms. Similarly, spaces dedicated to one (or a group of) function(s) and/or that require visual and/or acoustic privacy need to become separate rooms. For the rest, you can choose between open-plan – which maximises spaciousness – and compartmentalisation – which maximises mood distinctiveness. For this, it's best to start without preconceptions.

All these aspects of space affect well-being – hence health.

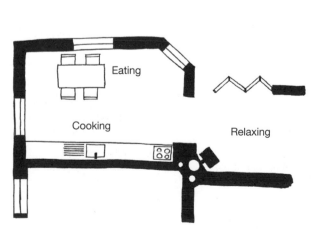

Functions requiring specific moods need to be distinct parts of rooms.

Alcoves or clearly distinguishable sub-spaces contain place- activity-moods.

Keypoints

- We need both physical and psychological space.
- Physical space should be efficient.
- Psychological space should match soul-need – which depends on activity and context.
- Spaciousness relieves stress.
- Light increases perceived spaciousness and lifts spirits.
- Every home needs a cosy social heart.

Choices

Options	Decision-making factors
Spacious and light-flooded or cosy, socially, thermally and spatially compact?	For what activity? Where? When?
	Outward-looking or protectively secure?
	Social cohesion or recuperative stress-relief?

Sensory nutrition

Our senses inform us about the world around us. They were formed long ago, when the world was so dangerous that survival depended on sensory warnings. They alert us to the need for action (e.g. running away from lions) or, if nothing needs attention, let us relax (so recover from exertion). Consequently, in today's world, minimalist simplicity and a limited palette of materials calms space, so feels like it expands it. However, as this purity suppresses the clutter life produces, over-minimalist décor can feel sterile, and calming restraint easily progresses to sensory starvation. Our senses need constant stimulation to stay awake. Without this, life is boring and we feel only half alive. Moreover, we need some traces of chaos as manifestations of life. But clutter clogs-up life, and unpredictability brings insecurity; this makes excessive stimulation stressful.

We need, therefore, *both* liveliness and calm, *both* stimulus and predictability: "difference within same-

ness.[11]" This can be subtle. Nature's gentle movements (e.g. waving grass, dancing-leaf shade-patterns, lapping water) provide this.[12] So do the slight irregularities of hand-made forms and textures. This tension between the awakening and the relaxing and their resolution in contrast-fusion applies to all our senses. Even with immobile buildings, sensory delight isn't just visual: aroma, tactility and warmth play a large part. Unpainted wood looks, feels and smells life-friendly. Plastic doesn't. Walls with sunlight-stroked textures invite hand and eye. Dead-smooth ones don't. Some things feel welcoming. Others feel hostile. Hard, sharp things hurt to bump into. Hard, smooth materials (e.g.

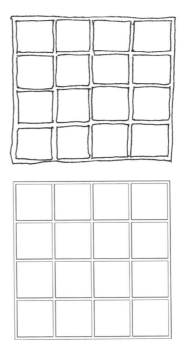

Tiling: hand-straight and dead-straight.

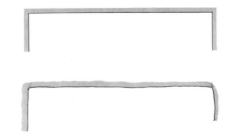

Dead-smooth and hand-smooth surfaces.

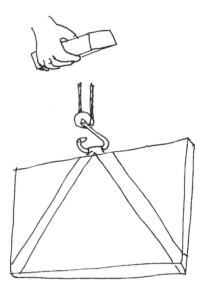

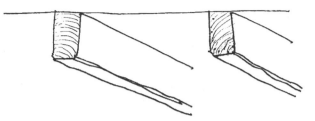

Beams with edges subtly shaved-off feel softer.

Bricks are objects at a human-scale – made to fit the human hand. Pre-manufactured panels are 'machine-scaled', so lack this imprint of 'life'.

steel or glass furniture, mirror-smooth walls and floors) repel our touch.

Besides being 'hard' on the spine, hard floors' tactile (and visual and acoustic) hardness resounds in the soul. Tiled floors are no softer than concrete or ter-razzo[†], but, being sized for hands not machines, they *look* softer – just as brickwork looks softer than con-crete. This effect is enhanced if colours are warm and

Conflict or conversation: as our eyes follow lines, these carry much power.

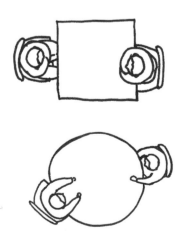

Round tables are less confrontational than square.

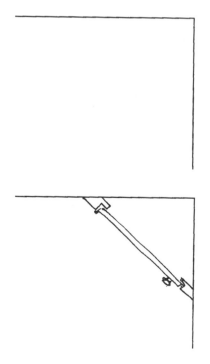

Moderating abruptness en-gentles eye-movements.

variegated and tiles aren't laid in *exactly* straight lines (e.g. laid without uniform spacers, and by eye, not to string-line. Fortuitously, imperfect tiles, bricks and pavers – 'seconds' – are cheaper than perfect ones.). Such imperfection, of course, isn't to everybody's taste. Nonetheless, obvious hand-laying of tiles or bricks imprints life. Mechanical 'perfection' has nothing to do with life.

Tactilely, visually and even socially, softening shapes helps. Tactilely, sharp edges hurt to collide with. Rounded-edge counters and shelves with edges subtly shaved-off feel softer, look gentler. The same for beams – and, as chamfered-edges ignite slower in fires, these are also traditional. Visually, whereas gentle shapes (e.g. arches, curved walls, rounded-off corners) are calming, sharpness activates the amygdala (the parts of the brain associated with fear and aggression).[13] Sharp shapes and the collisions between them induce staccato eye-movements, hence tension.[14] Where outlines, shapes and planes meet at right-angles, our eye-movements are particularly abrupt. Moderating such meetings (e.g. with corner furniture) en-gentle eye-movements and let the constituent elements 'converse' instead of fight. This affects our social predisposition. Similarly, round tables establish less confrontational relationships than square.

Related to this, we intuitively 'read' invisible structural forces – just as we intuitively know how hard to throw stones to reach a target. Curves are structural shapes for brick – as bricks are held together by gravity, not glued by mortar: only weight above them makes arches stable. Being linear, however, timber is better suited to faceted arches and polygonal room plans. Timber allows long horizontal openings to appear quite natural. Masonry doesn't: vertically proportioned openings make better structural, hence visual, sense.

In all such ways, for a soul-warming home, every form, shape and texture needs to feel caressed by hand, or its surrogate, the eye. And not just the eye: this applies to all our senses – of which we have many, not just five. How your home nourishes the senses and aesthetic sensibilities is a crucial part of imbuing it with soul.

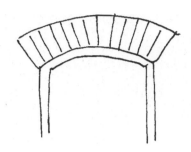
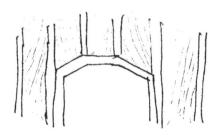

Arches for brick: faceting for timber.

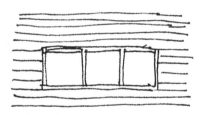

Long horizontal openings suit timber.

Masonry needs more vertical openings.

Keypoints

- We need both security and stimulus.
- Delightful multi-sensory ambience is soul-nourishing.
- Softer shapes/materials/textures are more welcoming than hard.
- Moderating right-angles en-gentles mood.
- Hand-sized components humanise buildings.

Choices

Options	Decision-making factors
Minimalist calm or sense-rich?	What are your soul-needs? When? Where?
Mechanically perfect, precise or hand-made?	Admiration for technology or delight in sensory nourishment and/or craft skill; functional efficiency; cost.
Soft or hard forms, shapes, textures and materials?	Dynamic drama or gentleness, approachability; safety; physiological health.

Soul/spirit factors

We are what we eat, but we're also *more than* what we eat. Both nature (genes) and nurture (environment) make us who we are. Environment includes *physical* environment: particularly home. How this is, how it sustains us and supports good moods, there-fore, matters a great deal. Health and well-being interact. Consequently, physical health can depend on our state of soul. Scientists call this effect 'psychoneuroimmunology'. It's certainly not worth building an energy-efficient home that you'll feel miserable in: misery leads to illness. (This isn't as unlikely as it might sound. To conserve energy, workplaces, hotel cubicles and even schools have been built without windows.)

A home's mood partly depends on what we do there, partly on what's physically (so sensorially) there and partly on the messages it emanates. The first thing that affects visitors and residents alike is the approach and entry journey. This is about how space contracts at gate, doorway and lobby, and expands into yard, garden and rooms; how *we* move through space and *its* movement gestures; what we sense and see; how these sequential experiences affect our moods; and what the things we 'meet' 'say' to us.

'Curb-appeal' influences first impressions – which colour subsequent impressions. Small buildings look less assertive, so are more welcoming than large ones. Minimising façade height – which confronts us – by having rooms in the roof, and tying buildings into place by shrubs at corners, helps this. (Such shrubs can also serve microclimatic functions; see: **Chapter 3: Choosing what should be where in the garden: microclimatic landscaping**, page 23.) Once indoors, just like whenever we sit in public, we feel most at ease with expansive views before us, and a wall at our back. This is a fundamental Feng Shui principle: 'prospect' in front, 'security' behind.[15]

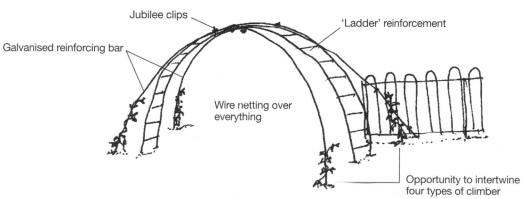

Galvanised reinforcing bar

Jubilee clips

'Ladder' reinforcement

Wire netting over everything

Opportunity to intertwine four types of climber

A green archway as entry-threshold. How to make one:

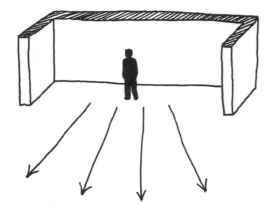

Prospect in front, security behind.

Minimising façade height reduces buildings' assertiveness.

Do you want your home to say it (and by implication, you) is focused on the future (the era of never-ending improvement) or the past (the golden age which will never be surpassed)? Unfortunately, futuristic design looks dated after a few years; and nostalgic design traps you into past patterns. Timelessness, however, synthesises forward-looking optimism and grounded stability. This starts with the materials you choose, both how they are now and how they'll age.

Indoors and out, construction materials influence what buildings say. Masonry conveys earth-bound protective stability. Glass is about open, light-filled levity. Timber is in-between security and levity but adds warmth. Fake materials (e.g. wood-grained laminates, glued-on brick-slips, wood-framed stucco) may *look* convincing but *feel* fake. Natural materials connect us to their source in living nature; industrial materials, to lifeless industrial processes. (This doesn't preclude the attractiveness of industrial-era buildings, like gasometers, mills, etc. It's just that steel pressings don't automatically make tables and chairs appealing to the soul. Wood and leather do.) Additionally, second-hand materials bring character imprinted by their former life use. They're also softened by age. Related to this, it's wise to choose materials that will be improved, not compromised, by aging.

Buildings are more than just materials, though. Their form gives them meaning. Their proportions also affect us. Horizontality is calming; verticality asserts its presence. Proportions can also manifest cosmic harmonies. Hence the Golden Mean ratio (1:1.618…), which orders both human bodily proportions and

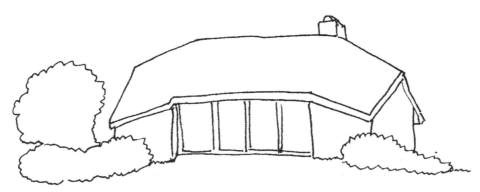

Tying buildings into place by shrubs at corners reduces perceived scale.

What is spirit-of-place? What is soul-of-place?[16]		Core concept
Spirit-of-place	Embodied values, manifest in transformative beauty or crippling ugliness. It's invisible and intangible, but raises or crushes our spirit.	Truth (integrity), goodness (altruism-motivated), beauty (transformative)
Soul-of-place	The place's mood (produced by both physical environment and activities) that affects our moods. It's experienced through our (many) senses, and nourishes or poisons the soul.	Multi-sensory and social/emotional mood
These relate to and nourish – or assault – our human spirit and soul: spirit being the core of us, manifest by the values we live by; soul, our baseline mood, which affects how we relate to the world.		

cosmic spatial relationships, induces harmony; and 1:1, although often lifelessly static, is about balance. It's particularly *how* they're made that gives buildings soul – or robs them of it – and the values that their design embodies affect the spirit they emanate. However attractive the design, if there's no handwork in its making, no product can emanate soul. Machines aren't alive, can't feel and have no soul: so can't impart it. For factory-built houses, therefore, the ensouling process can only start when occupants live in and love them. With hand-made things, however, soul is embodied through their making process. As hand is connected to heart, we can't help putting heart-forces into anything we make.[17]

However, although you can't hand-make anything without engaging your artistic sensibilities, this isn't about making things look 'arty'. That just makes them feel contrived. Things shaped by many layers of purpose (e.g. climate-response, energy conservation, functionality, structure, mood-support, aesthetic harmony) have integrity. Integrity lets us trust our surroundings: essential for inner peace. This is something we all sense; it doesn't need thought. We *sense*, not think, a place's spirit.

The soul-warm resonations of human social life give more soul to a place than any lifeless physical thing, and can make even ugly buildings homely – but this is an uphill process. It helps, therefore, if soul and spirit are already embodied in your home by its making process – especially that which you yourself have imparted. This, however, takes time, so is usually expensive. Affordability, therefore, often limits hand-made things to the parts of houses most frequently seen and touched. Consequently, the more that you

can do yourself, the more soul will be embodied in your home. The resulting soul-nourishment normally outweighs any lack of technical perfection. Doing things yourself also saves money: finishes (and fittings) – which don't destroy the house if done badly – account for some third of its cost. This is about *process*. In this respect, design is only a part of process.

Keypoints

- Externally, small perceived scale feels more welcoming than large.
- Natural materials connect us to life; industrial materials to industrial processes; fake materials exude a niggling feeling of dishonesty.
- Choose materials that improve with age.
- Hand-work imprints life and embodies soul.
- For integrity, aesthetics need to be shaped by many layers of purpose.

Choices

Options	Decision-making factors
Message: life-connected or machine-made?	Comparative value accorded to technology or life; functional efficiency; embodied soul and spirit; cost.
Style: high-technology, traditional or timelessly style-less?	Future-inspired, nostalgic or focused on the eternal present.

Nature connection

There's also another soul nutrient of a completely different kind: nature. Humanity was formed in fertile but unprocessed, 'untamed' nature. In an evolutionary scale, agriculture is a recent phenomenon; built environment, very recent; and lack of immediate access to greenery barely a few generations old. We retain strong archetypal associations with this (e.g. water,

 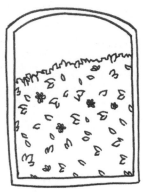

Wall view: hedge view.

and, particularly, greenery) and also modifications that we depended on (e.g. ensocialising fire, security). Consequently, all the ways our environment differs from a nature-formed and archetypal one affect us. The effects of such divergences aren't always adverse, but always are important to consider.

Living nature confers life energy to us. If bed-bound, we can look at breeze-stirred foliage for hours; bare walls are boring after a few minutes. Greenery counters the aridity of solely *built* surroundings. Even indoor plants relieve stress.[18] (Indeed, finding that green views reduced hospital patient recovery times by 21%, Professor Roger Ulrich considers every tree-leaf visible from a window "worth its weight in gold."[19]) Greenery, in fact, is all about life, so re-enlivens us. Indeed, it's a potential habitat for many levels of life: from lichen and moss to insects and birds. It moderates microclimate and drinks up groundwater (or requires watering). Vegetation, however, grows as it wishes – not always how we wish – so needs maintenance (e.g. mowing, pruning, weeding).

Maximising green view with diagonal axis. Obstructions increase perceived extent.

Mirrors in corners can treble both the light plants get and the greenery we see.

Even the tiniest plots have more opportunities for greenery than is generally supposed: trees, wall surfaces, green archways, ground, roofs and pots and planters. Similarly, for stress-relief, windows should focus on as long and as green views as possible. The more greenery in sight and more spacious our surroundings, the more relaxed we feel.

Keypoints

- Humanity was formed in, so is adapted to, a natural environment.
- Greenery is relaxing and re-enlivens us.
- It supports other levels of life.
- Maximise green views and perceived extent of green surroundings.

Choices

Options		Decision-making factors
Hard or living surface?	Ground	Foot/wheeled traffic; mowing, weeding; stormwater run-off, percolation; adequate irrigation water? Microclimate; aroma; appearance; appropriateness of message.
	Walls	Wall foundations/stability; view; blossom aroma; songbird/insect habitat; intruder access/deterrence; pruning access/ease.
	Roof	Skyline; songbird/insect habitat; stormwater run-off; adequate irrigation water?

Resources

Information

Alexander, C., Ishikawa, S. and Silverstein, M. (1977) *A Pattern Language*, Oxford University Press, New York.

Day, C. (1998) *A Haven for Childhood*, Starborn Books, Clunderwen, Wales.

Day, C. (2007) *Environment and Children*, Architectural Press, Oxford.

Day, Christopher (3rd ed. 2014) *Places of the Soul*, Routledge, London.

Day, C. (2002) *Spirit & Place*, Architectural Press, Oxford.

Hobday, R. (2000) *The Healing Sun: Sunlight and Health in the 21st Century*, Findhorn Press, Forres, Scotland.

Olds, A. R. (2001) *Child Care Design Guide*, McGraw Hill, New York.

Swire, S. (2015) *Secrets of the Feel-good Home: What Works and What Doesn't,* Wordzworth Publishing, Tunbridge Wells.

Venolia, C. (1988) *Healing Environments,* Celestial Arts, Berkely, CA.

1. Cancers, organ failures, cardio–vascular and neurological diseases and traffic accidents are major killers.

2. *www.dirtyelectricity.ca/electrical_health_implications.htm* (accessed: 22 May 2014).

3. Ibid.; Hawkins, L. (1989) Health problems arising from geopathic stress and electro-stress, paper given at Sick Buildings and Healthy Houses seminar, London 22–23 April 1989.

4. An infra-red study of Regensburg showed medieval streets followed the lines of subterranean water courses, thereby ensuring houses avoided them: Eble, J. (1981) Unpublished lecture, Järna, Sweden.

5. Bachler, K. (1989) *Earth Radiation,* Wordmasters, Manchester.

6. Does a tree die when felled? Willow stakes can sprout many years after they were cut. Do humans die when the heart stops, brain damage is irreparable, hair stops growing or decomposition starts? When is the moment life ceases?

7. Gapell, M. Sensual interior design in building with nature; Daniels, R. (1999) Depression – a healing approach, *New View*, 1999, 4th quarter; Venolia, C. (1988) *Healing Environments*, Celestial Arts, Berkely, CA; Hobday, R. (2000) The healing sun, *Building for a Future*, Summer 2000, Vol. 10, No. 1.

8. *http://ergonomics.about.com/od/kitchen/f/work_triangle.htm* (accessed: 23 December 2013).

9. Laminated glass is equally safe, but obstructs solar heat.

10. This sensory aspect of warmth isn't at all insignificant. Many Polish migrant workers keep 'warm' at night in unheated British houses by burning a night-light under an inverted flowerpot. The actual heat contribution is minute, but the warm, flickering light makes them feel warmer.

11. Fiske, D. W. and Maddi, S. R. (1961) Functions of varied experience in Olds, A. R. (2001) *Child Care Design Guide*, McGraw Hill, New York.

12. Kaplan, S. (1995) The restorative benefits of nature: toward an integrative framework. *Journal of Environmental Psychology*, Vol. 15, and Kaplan, R. and S. (1989) *The Experience of Nature: A Psychological Perspective*, CUP, Cambridge, in: Nute, K. (2006) The Architecture of Here and Now: Natural Change in Built Spaces (proposal for The Architectural Press, Oxford).

13. Bar, M. and Neta, M. (2006) Humans prefer curved visual objects, *Pyschological Science,* 2006, Vo.17; Lidwell, W., Holden, K. and Butler, J. (2007) *Universal Principles of Design*, Rockport Publishers (Quayside Publishing)

Beverly, MA, in Swire, S (2015) *Secrets of the Feel-good Home: What Works and What Doesn't*, Wordzworth Publishing, Tunbridge Wells.

14. On the other hand, clear geometric shapes can be reassuring in times of stress. These qualities aren't necessarily incompatible.

15. Ibid.

16. These are elusively ambiguous – and sometimes interchangeably used – terms. This, however, is what I mean by them.

17. Don't tradesmen also do this? Rarely. They've been trained to produce uniform machine-standard 'perfection', and do it fast: no heart-forces.

18. Kay-Williams, S. (2007) Don't forget – the garden, *Green Building* magazine, Summer 2007, Col. 17, No. 1; Indoor plants increase productivity and well-being: Fjeld, T. (study in a Norwegian oil company's offices); Wendt, A. (2008) Bringing nature indoors, *Environmental Building News*, October 2008, Vol. 17, No. 10.

19. Professor Roger Ulrich, University of San Antonio, Texas, BBC interview 2002.

Keeping the home safe

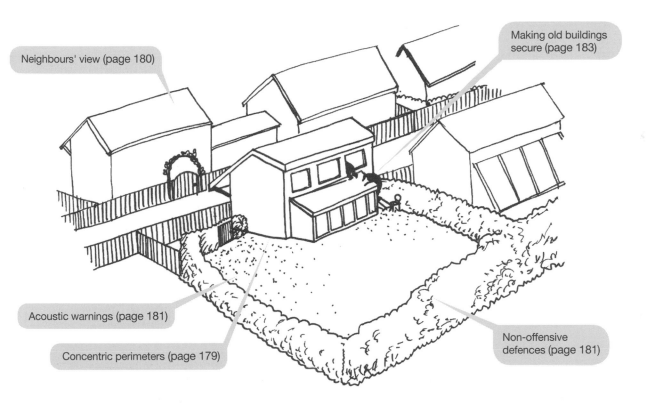

CHAPTER FIFTEEN

Neighbours' view (page 180)

Making old buildings secure (page 183)

Acoustic warnings (page 181)

Concentric perimeters (page 179)

Non-offensive defences (page 181)

Security basics

Constant tension stresses health, but we can't be relaxed if we feel insecure. To improve security, we tend to first think of locks on doors and windows. But locks should be the *last* line of defence. The first one is keeping burglars from *wanting* to try to break in. This requires concentric perimeters. Just as for keeping animals *in*, concentric perimeters are effective for keeping burglars *out*.

The first perimeter is the community boundary. Within this, it should be clear who owns what, and that the whole community has an interest in its security. The more that places are in neighbours' view, the more obvious it is to potential burglars that their presence and any suspect behaviour will be noticed. Most bur-

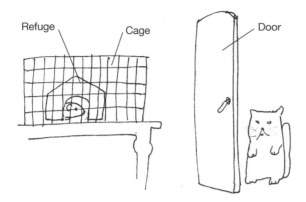

Concentric security perimeters: keeping cats from pet hamsters.

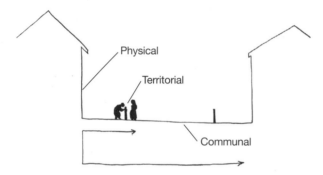

Concentric security perimeters: houses.

Portal-arches and bridges provide clear realm markers for communal closes.

Keypoints

- Establish concentric perimeters.
- Within the first (community) perimeter, it should be clear who owns (or is responsible for) what.
- Communal closes need clear realm markers.
- The more that places are in neighbours' view, the more do potential burglars feel their presence and behaviour will be noticed.
- Psychological messages also deter burglars.
- Upkeep is important.
- Windows and doors must be secure.

glars are opportunists, so are further deterred by psychological messages that emphasise territoriality (e.g. portal-arches, bridges, different ground surface) and propriatorial care (e.g. upkeep, cared-for gardens, window boxes).

The next perimeter is the territorial boundary. Territory markers (e.g. gateways, threshold-marking paving) clarify its extent. Within it, uninvited visitors are obviously trespassers. This also should either be under the surveillance of neighbours and/or passers-by; or it must be protected by an *un-climbable* barrier.

The third perimeter is the house exterior: its walls, windows and doors. These, of course, must be secure. This, however, is the 'last resort' perimeter. The prime security comes from the community around us.

Burglar-proofing

Besides obvious security measures, there are some that aren't always thought about. A major one – as should be clear by now – is unobstructed visibility by others: overlapping arcs of view. As burglars don't enjoy being seen at work, houses in public sight get burgled less often than secluded ones. Anywhere that burglars could climb in, therefore, should be fully in neighbours' view. The most at-risk part of properties, however, is the private part – usually the rear – the bit

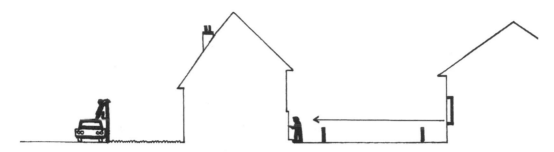

The front of your property should be in neighbours' view, while the rear should be completely inaccessible to strangers.

Climbable fences mustn't obscure neighbours' view.

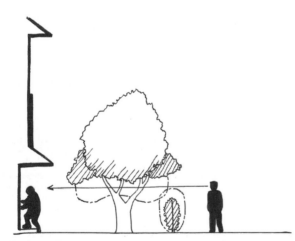

Shrubs and trees should be pruned to keep doors and windows visible.

Kitchen window supervising the front door.

you *don't* want anyone looking into. This means front and rear need different security treatment.

At the front, if strangers can access the house, there should be clear realm markers to make their presence noteworthy. Similarly, any climbable fence must *not* obscure neighbours' view; nor foliage hide weak points, like doors and windows. (You may need to prune trees and shrubs for clear viewlines.) View from your own windows is also important. One day-room window should supervise the gate (which should clink, creak or tinkle a bell) and the area around the front door. For night time, there can be other auditory warnings (e.g. deck-boards fixed loosely enough to creak, and gravel paths that crunch). These also deter burglars by making them aware that they might be heard.

As the sides and rear of the house are out of sight, these should be completely inaccessible. Out-of-sight windows are particularly high-risk. These should be opening-restricted or (if they can be safely accessed for cleaning from the outside) non-opening. You should also be careful that any extensions, garage roofs, vines, trees (or parked vehicles) don't give access to upstairs windows. Deeply overhanging eaves, no close trees and no movable objects (e.g. garden furniture, dustbins, etc.) to use as step-ups make roofs hard to climb on to.

You may also need to raise fences and/or clad them in roses or thorny plants. Although formidable barriers, roses are attractive landscaping and ecological assets, so 'non-offensive defences'. Non-offensive barriers can also be used outside vulnerable ground-floor windows. For example, a water pool, shallow enough for child safety, or prickly thistles and shrubs, which merely bring irritating discomfort. These won't

Deep overhangs make roofs hard to climb on to, especially if they're steep and/or have slippery surfaces.

Rose / thorn-clad fences are hard to climb over.

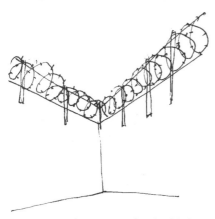

Secure. But who wants to live in this?

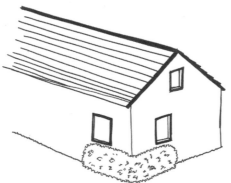

Thorns outside vulnerable windows – but not barbs: windows are fire escapes!

Roses are less offensive than razor wire.

stop a determined intruder but do act as deterrents. (But remember that windows are fire escapes, so thorns may hurt, but mustn't *grip* clothing.) One drawback with thorny bushes is that they collect air-borne polythene bags. Shrub breadth, therefore, shouldn't exceed grab-stick reach, so you can remove such litter.

As it's our surrounding community that best assures security, our security measures should reciprocate by conferring environmental benefits. Whereas razor wire says: "We're enemies", so breeds mistrust, non-offensive defences say: "We're all friends here." By giving something to the community, non-offensive defences both encourage community good-feeling *and* increase security. And they look and smell nice and support songbirds and wildlife.

Keypoints

- Ensure front boundaries are in neighbours' view; rear ones, inaccessible to strangers. Prune shrubs and trees to maximise door/window visibility.
- Any boundary not surveilled by neighbours and/or passers-by must be un-climbable.
- Climbable fences mustn't obscure neighbours' view.
- Use roses and/or thorns (which may be barbed) on fences; and (non-barbed) non-offensive defences outside vulnerable windows.
- Locate windows for good supervisory views.
- Use auditory warnings.
- Deny access to upstairs windows.
- Ensure nothing growing, built or movable can assist climbing over fences or into the house.

Choices

Options	Decision-making factors
Community membership or anti-social insularity?	Privacy; strength of community; neighbours' trustworthiness.
Boundary markers or physical barriers?	Communality; children's social autonomy; neighbours' trustworthiness.
Opaque wall/hedge or transparent fence?	Climbability; neighbour's surveillance.
Razor wire or roses?	Trust-building; message; amenity; environmental quality; maintenance; litter removal.

Making an old building secure

For security, the same considerations apply to old buildings as to new ones. Old buildings tend to come with an established community: usually an advantage. Their locks, however, are likely to be less secure (e.g. can be opened with a credit card), so may need replacing. Likewise, sash window latches are easy to slide open from outside, so the sashes need locking fitches or drive-screws to secure them together, which also tightens them against draughts. (Hang the key close by, in case you need to use the window as a fire escape.) Alternatively, you can drill a hole (0.5-1mm oversize and sloping slightly down) and slip in a waxed 5" (130mm) nail, cut to length. This is faster to remove in case of fire, and cheaper, but doesn't clamp top and bottom sashes together, so you need other measures (e.g. brush-strips) to draught-proof them.

Like a front-door chain, it's also advantageous to fit opening restrictors on all accessible windows. So long as they're out of grab-stick reach from each other, this lets you keep windows ajar for cross-ventilation cooling at night and while you're out. With hotter summers, this is likely to become an increasingly frequent necessity.

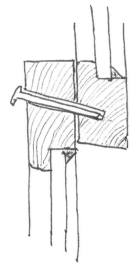

For sash windows, you can drill a slightly oversized hole (sloping down) and slip in a waxed 5"/130mm nail, cut to length.

Keypoints

- Apply the same security measures to old buildings as to new ones.
- Replace easily forced locks.
- Secure sash windows.
- Opening restrictors on windows ensure security while allowing cooling ventilation at all times.

Choices

Options	Decision-making factors
Manufactured or improvised sash window locks?	Fire-exit convenience; security; draught-proofing; cost.

Resources

Newman, O. (1973) *Defensible Space*, Collier Books, New York.
Stollard, P. (1991) *Crime prevention through housing design*, Spon, London.
www.dailymail.co.uk/news/article-2107511/The-home-guard-Police-suggest-30-thorny-bushes-homeowners-plant-discourage-lazy-garden-thieves. html

Whole-life access

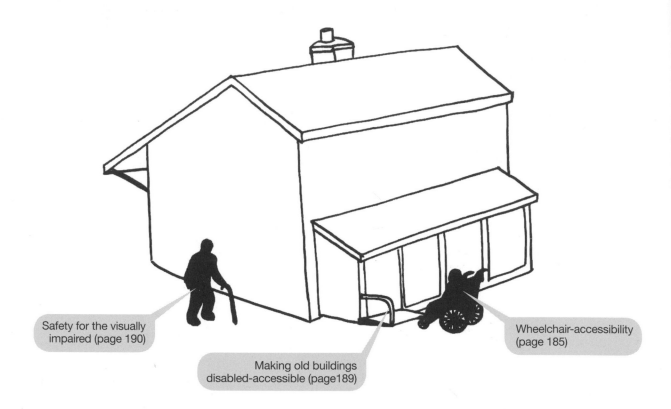

Safety for the visually
impaired (page 190)

Making old buildings
disabled-accessible (page189)

Wheelchair-accessibility
(page 185)

Disabled access basics

All of us age – or, at least, hope to live long lives: different words for the same thing. If you build something you enjoy living in, you probably want to live there when you're elderly. Also, if you have elderly relatives, you'll want to ensure your house is accessible for them – you don't want to prevent grandparents from visiting grandchildren.

With age, our mobility, eyesight and hearing gradually decline. For such infirmities, this means disabled-accessible design in one form or another. Not all disabled people use wheelchairs: most are just unsteady. Nonetheless, it's wise to allow for them: a single accident or cardiac event (at any age) can turn an ambulant person into a wheelchair-bound one. Wheelchairs need space to turn, and don't like steps. All ground-floor floors, therefore, should be level, with neither trip nor slip hazards. At least one entry should be step-free.

Wheelchair-accessible light switch?

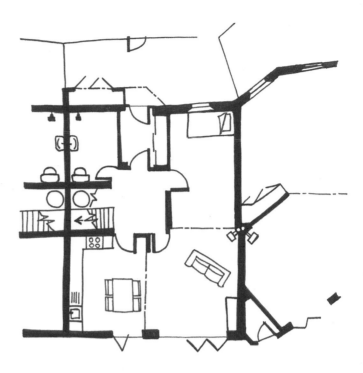

Kitchen / dining room, wetroom / toilet and (potential) bedroom on the ground floor.

Disability means non-ability – and, as there are thousands of abilities we take for granted, there are thousands of forms of disability. In fact, every individual's disability is individual, unique. This chapter can only generalise about common forms. The measures described will (more or less) suit most – not all – disabled people, but may need to be tweaked for individual need. As there are lots of things we tend not to think about, like having light switches in reach (e.g. at about 1m/3'4" off the ground, or pull-switches), it's not a bad idea, therefore, to rent a wheelchair for a day and list all the things you *can't* do.

Keypoints

- Build for the future: mobility and eyesight usually decline with age.
- Elderly relatives may soon have mobility difficulties.
- Keep ground-floors step-free and non-slip.

Physical accessibility

To build a house accessible for ourselves when we're elderly and infirm and, before that, for our children's grandparents, needs forethought from the outset. Retrofitting lifts is expensive and space demanding. Lifts require access and egress space; and wheel-chair stair-lifts require manoeuvring space at the top and bottom of the stairs. Moreover, both lifts and stair-lifts depend on electricity – which, in an oil-depleted future, can't be assured. New houses, therefore, should include, on the ground floor, some 'core rooms'. At the very least, these should include a kitchen to eat in, a wetroom/toilet and a room convertible to a bedroom.

Even for short visits, accessible toilets are essential. If there's space to turn a wheelchair, toilet rooms only need a shower fitting and floor-drain to become wetrooms. These should be 1.6 x 1.85m, 1.5 x 2m or larger,[1] *not* have lipped or raised shower trays but be completely flush-floored, and have shower-curtains or across-room screens instead of narrow doors. (Or you can sink the shower tray and top it with duck-boards.) You should allow for handrails in front of the toilet and in the shower; also one over the washbasin. Handrails or, more economically, scaffold-tubes along all walls (at around 1m/3'4" height), are a simple way of ensuring all-round support. Any handrails and fittings anywhere (e.g. washbasins, shelves, towel-rails, radiators), which people can use to steady themselves, *must* be as secure as they look, so need solid walls to fix to. (For this, plasterboard, hempcrete and straw-bales aren't strong enough, so need appropriately placed structure. Nor are most kinds of insulation block adequately strong.)

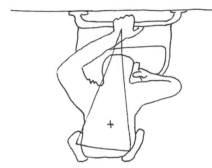

Stable: this keeps centre-of-gravity within three-point support: one hand and two feet.

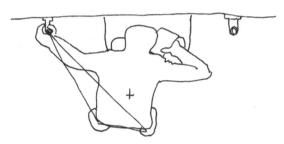

Unstable: handrails to both sides.

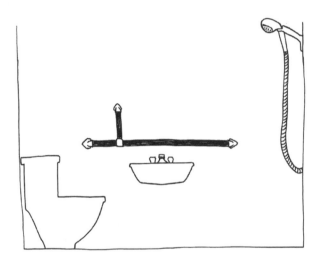

Flush-floored wetrooms.

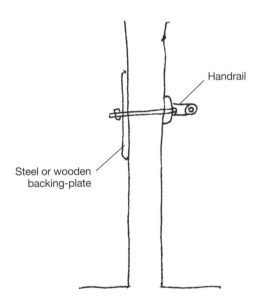

Handrail

Steel or wooden backing-plate

For fixing handrails to insulation-block walls, reinforcing or load-spreading backing-plates in the next room may be necessary. (These can be concealed within built-in furniture.)

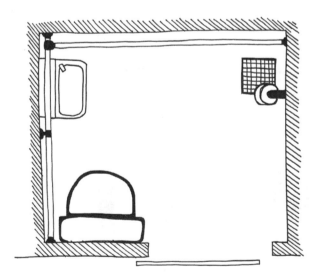

Wetrooms: handrails (e.g. scaffold-tube) all round, with wide-opening doorway.

Although wheelchairs can surmount 15mm (5/8") thresholds – with some difficulty for the infirm – such steps can trip the poor-sighted; although legal,[2] they're best avoided. Instead of (or as well as) steps, there should be ramps, no steeper than 1:12 (8%). If steeper than 1:18 (6%), or likely to become icy in winter, these need handrails. For stick- and crutch-users, all floors should be non-slip: in wetrooms, even when wet and soapy. Small mats that can slide on polished (or gloss-varnished) floors are particularly hazardous. Similarly, outdoor ramps and decks need ribbed-planking textures or brick-paving joints *across* direction of travel. Wheelchairs and Zimmer frames need space. Standard doors (860mm/34" frames) are *just* possible to wheel through but need perfect alignment and knuckles insensitive to pain. A 1m (3'4") doorframe is much easier. Similarly, passages should be 1m wide. As wheelchairs' outer turning radius is about 750mm (30"), wider passages and wider or recessed doorways ease turning in to rooms. (1.2m/4' passages give space for shelves if there's no immediate need for wheelchair access.) Better than very wide passages are widenings for 1.5m (5') turning-circles (or 1,400 x 1,700mm / 55 x 67"

1.5m (5') wheelchair turning-circle by core-room doors.

ellipses) outside the entrance to core rooms (kitchen, wet room, bedroom and preferably living room) and also within those rooms. As wheelchairs only need little headroom, part of this can be under the stairs (as long as there's room for wheelchair-pushers). This makes three things vying for space in the centre of the

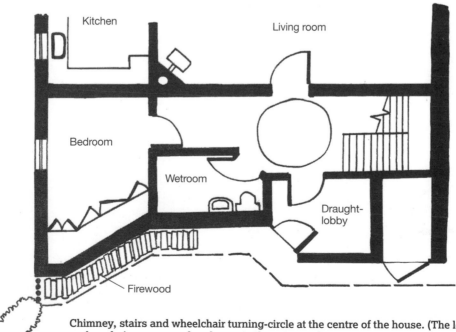

Chimney, stairs and wheelchair turning-circle at the centre of the house. (The lower eaves at each end give a sense of enclosure to the entrance.)

Powered wheelchairs are powerful. It's unwise to skimp on space.

house: chimney (or accumulator) for heat distribution; stairs to minimise circulation space (which is expensive to build and heat); and turning-circles for wheelchair accessibility. Fortunately, as 'the centre' isn't a single spot but the middle third under the roof ridge, these aren't impossible to reconcile.

These measures allow disabled relatives and friends to visit, but if you become disabled, it's likely that you won't be able to use your upstairs rooms, or clean them. Instead, you could let them as an apartment.

Besides the downstairs wetroom/toilet, there's usually a bathroom upstairs – hence also plumbing connections for a future kitchen. However, there are also fire-separation and noise-proofing issues. A little forethought can ease these: the entry passage needs to be wide enough to accommodate a dense-block one-hour fire (and acoustic) partition to wall-off the stairs

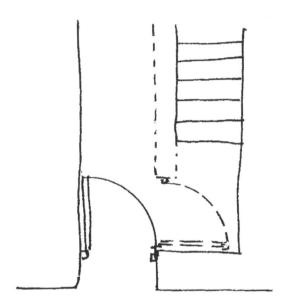

It only takes a little extra space to have the option of fire-separating upstairs and downstairs, so that you can let the upstairs.

The Lifetime Homes Design Guide: a summary[3]

- Easily accessible and wheelchair-friendly parking and access to building.

- Minimum hallway width: 900mm (36"), 1200mm (48") preferred.

- Minimum (clear) door width: 750mm (30") if head-on approach, otherwise up to 900mm (36") in narrow corridors. All doorways with a 300mm (12") wide space (e.g. a 300mm nib of wall) clear of any obstruction by the pull-side handle.

- Manoeuvring space a clear 1,500mm (60") circle or 1,400 x 1,700mm (55 x 67") ellipse, in wetroom and circulation spaces.

- Core living-facilities, including a (at least temporary) bed-space on ground floor.

- All wetroom/WC walls should support grab rails. Lever taps to washbasin.

- Stairs: 900mm (36") clear width, or opportunity to install a 1,000mm x 1,500mm (40 x 60") (minimum) lift.

- Potential to install a hoist in bedroom and bathroom.

- Windows: unobstructed view 800-1,700mm (32-67")above floor level. Handles/controls: not above 1,200mm (48").

- Switches, sockets, heating controls, etc. to be within a 450-1,200mm (18-48") height band above floor and at least 300mm (12") from room corners.

- Provision of fused spur cabling to aid potential future adaptations (e.g. stair-lift, platform-lift).

(allow at least 150mm/6') and long enough for a fire door and door swing-free space (at least 1,160mm: 45¾') at the staircase foot. It's also wise to arrange electricity and heating circuits so they can be separated off if so required.

The other thing to allow space for is disabled aids. These are surprisingly bulky. Powered wheelchairs are big. Hospital beds, with space for them to tip and hinge, are some 300mm (12") longer than normal beds. Recliner armchairs need almost a metre behind them so reclining doesn't destroy other furniture. And there's lots of intermittently used stuff (including wheelchairs) to find a home for. In many modern houses, rooms are too small. Compaction makes good sense for energy conservation (and developers' profit). For disabled residents, though, it can make homes unusable. Being more about critical dimensions in key places than spaciousness, however, disabled-accessibility needn't greatly increase house size.

There are regulations about disabled accessibility in public buildings, and will soon be (less demanding) ones for homes. Much more important than regulations or 'lifetime homes' criteria, however, is disability awareness. Mindlessly applied regulations hardly help. Awareness can make non-regulation spaces tolerably accessible.

Choices

Options	Decision-making factors
Ground-floor core rooms, space to install a lift or staircase easy for stair-lift installation?	Footprint; top and bottom staircase landings; staircase layout; plan for lift retrofit contingency; time required for retrofit.
Whole house adaptable to disabled access or ground floor already accessible?	Preparedness for sudden need; space available; time required for retrofit; attractiveness or indispensability of upper floor(s).

Making an old building accessible

Although the same considerations apply to old houses as to new ones, it can be more demanding to implement them within a structure and layout that has already been built. The first priority is a disabled-accessible entrance. The unsteady need handrails or handgrips beside doorsteps. For wheelchair-accessibility, you need to replace steps with sloping paving to a 4' (1.2m) level landing by the door (as required by regulations); or – less trouble to fit but more to use – just lay a portable ramp over the doorstep. As doorsteps are threshold-markers, you can instead devise a sensory (e.g. visual, textural, acoustic) territory-marker experience to emphasise the public-to-private transition.

Next come accessible core rooms. At the minimum, these must include somewhere to eat and sleep – for which it's usually easy to adapt ground-floor rooms – and toilet facilities. It's usually possible to find somewhere (e.g. under the stairs) for a small toilet for the ambulant disabled – but not to find one big enough to turn a wheelchair. A proper wetroom may necessitate building an extension – or fitting a lift or stair-lift and adapting the bathroom. If these options are impracticable or unaffordable, though, you may have to resort to barely tolerable improvisations, like leaving toilet doors open for manoeuvring space, washing in kitchens and toileting in commodes. The better the provi-

Keypoints

- Ensure ground-floor core rooms (kitchen / dining room, wetroom/toilet, potential bedroom and optimally living room) are accessible to the disabled.
- Use ramps instead of steps / stepped thresholds.
- Wetrooms should be flush-floored.
- Ensure relevant walls are solid enough for handrails.
- For wheelchair access you need minimum 1m (3'4") wide passages and (if entry isn't head on) 1m door-frames, and 1.5m (5') turning circles (or 1,400x1,700mm/55 x 67" ellipses) within core-rooms and by their doorways.
- Make provision for separating off the upstairs as a self-contained apartment.

sions, however, the easier and less dignity compromising for all concerned.

Most houses were built before disability regulations were thought of, so few homes could conform to these regulations were they obligatory. With enough thought, however, most houses can become – at least tolerably – disabled-accessible.

Keypoints

- If no space for a wetroom, a small ground-floor toilet is better than nothing.
- Or build an extension.
- If neither is possible, improvise, but do your utmost to provide for all essential activities.

Choices

Options	Decision-making factors
Wetroom or a small ground-floor toilet?	Ambulant-disabled or wheelchair-bound relative/friend; deteriorating or stable mobility; duration/frequency of disabled person's visit; opportunity to subsequently build an extension, fit a lift or stairlift.
Permanent or portable ramp?	Heritage building; frequency of disabled person's visit; constructional implications; space; cost; storage of portable ramp.
Fully or just tolerably disabled-accessible?	Constraints of existing building; frequency of disabled-access need; severity of disability; tolerance to improvisation.

Safety for the visually impaired

Many elderly are fully ambulant but visually impaired. They therefore can use stairs, steps and short-cuts inaccessible to the wheelchair-bound. In your own house, you know where hazards are, even in the dark. Visitors, however, don't. For them, any potential hazards (e.g. steps) and things they need to notice (e.g. doors, door-handles) should be conspicuous. This means textural, colour and, especially, *tone* distinction, and preferably contrast, from their surroundings. Also, ensure that lighting clarifies form by how it casts shadows. In particular, avoid flat lighting (e.g. from fluorescent strip lights) from behind you, as this can flatten-out the appearance of steps, so you don't know they're there. As we know our own homes, it's easy to forget that others don't. This is a critical safety issue.

Keypoints

- Any potential hazards need high visibility.
- Contrasting tones and colours increase visibility for the visually impaired. Different textures also help.
- Lighting should clarify form.

Choices

Options	Decision-making factors
Direction of daylight and artificial light?	Visual clarity of hazard; form modelling by shade; obscuration by cast shadows.
Diffuse or focused electric lighting?	
Change of materials?	Tonal contrast; contrast when dirty; cleanability of distinctive textures.

Resources

Information

Building regulations, part m, HMSO.
Day, C. (2010) *Dying: or Learning to Live*, Trafford
Publishing, Bloomington, IN
www.lifetimehomes.org.uk

Stair, platform and corner lifts

www.abilitylifts.co.uk
www.stannahlifts.co.uk
www.terrylifts.co.uk/products/domestic-wheelchair-lifts/s7-xpress-wheelchair-stairlift/
www.tkencasa.co.uk
(among many others)

1. My own is 1.5 x 1.65m: just usable, but rather awkward.
2. In Britain, as are all the regulatory requirements that I cite.
3. *www.lifetimehomes.org.uk/pages/quick-print-version-revised-criteria.html*; the Lifetime Homes Design Guide, Lifetime Homes, Holyer House, 20-21 Red Lion Court, London.

IV. WHAT ISSUES

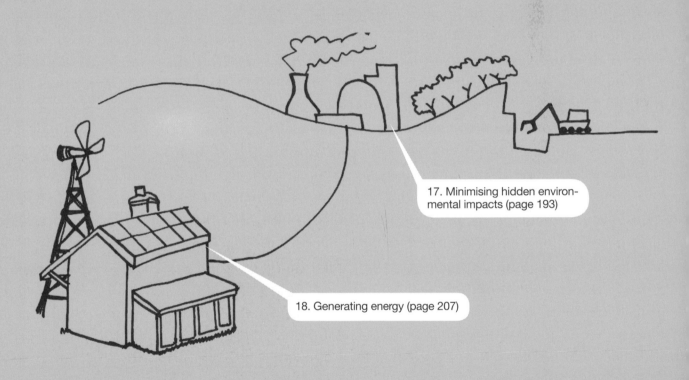

Minimising hidden environmental impacts

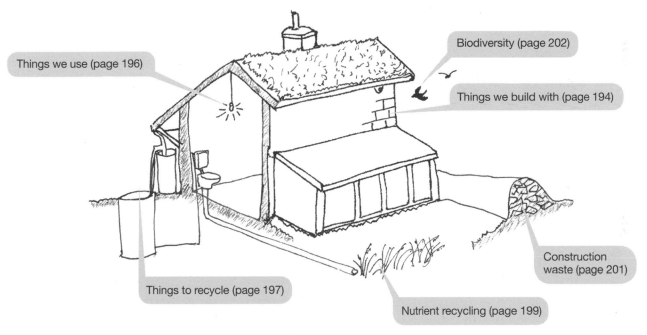

CHAPTER SEVENTEEN

Things we use (page 196)

Biodiversity (page 202)

Things we build with (page 194)

Construction waste (page 201)

Things to recycle (page 197)

Nutrient recycling (page 199)

Environmental impacts basics

There are three critical aspects to environmental impact: resource depletion, pollution and ecological damage. Resource depletion impoverishes future generations. Pollution – especially CO_2[1] – threatens life, *all* life. Ecological damage has unknown consequences, potentially as serious.

For life to survive, all adverse impacts must be minimised. It may seem that these impacts are all caused by other people, particularly industrialists. However, not only do *we* buy their products, but built environment accounts for around half of all environmental damage – and a large part of this is CO_2 due to homes. In fact, *we* – as householders – can easily greatly reduce adverse environmental impacts simply by making careful choices in three areas:

- How we heat and cool our homes.
- The materials we build with.
- The things we use and get rid of – and how we do this, particularly with awareness of the relationships and eco-systems these affect.

Heating and cooling have been covered in previous chapters (see: **Chapters 6-8 and 10-12**). This chapter discusses impacts embodied in materials and those due to things we use, and how we use and dispose of them. CO_2 reduction involves both daily energy consumption and embodied energy in materials. Reducing other pollution involves choosing low-toxicity materials and production processes; and closing cycles so we dump less waste. Reducing habitat destruction and flooding requires understanding the consequences of our actions and acting accordingly. Most adverse impacts are invisible at a single-building scale, but together, they threaten planetary life. Many, however, are wholly avoidable.

Keypoints

- All buildings cause resource depletion, pollution and ecological damage.
- All such adverse impacts can be greatly minimised.
- Impact minimisation includes how and of what we build buildings, what we use in them, how we dispose of what we don't want – and awareness.

Materials we build with

Reducing CO_2 pollution is imperative and urgent. With about a half of this being due to buildings,[2] this particularly involves reducing the energy they use – or, more exactly, the energy *we* use in them. Bills make us aware of the energy consumption due to operating buildings (mostly heating and cooling them). But the energy – hence CO_2 – embodied in their materials and construction, although only averaging a third of life-time operating energy,[3] is no less serious. Whereas operating CO_2 is released over a building's lifetime, manufacturing CO_2 is released within months, effectively doubling its climate impact.[4] The longer buildings or materials reclaimed from them last, however, the longer period embodied energy is amortised over, effectively reducing its impact. Re-used materials are cheaper but usually require more labour, hence – if you employ contractors – increase cost (but see: **Chapter 19: ou build your house, page 223**).

Most manufactured materials have high embodied energy and embodied pollution. Long-lasting materials (e.g. brick, concrete), however, somewhat offset this. Natural materials – including newly revived ones (e.g. unfired clay, straw) and composite developments (e.g. hempcrete, wood-wool) – have extremely low embodied pollution. Cellulose materials anchor carbon, so are 'carbon-negative'. (Only some timber species, however, survive long outdoors in damp climates and many require repeated maintenance, e.g. painting, oiling.) Also, *local* natural materials embody little transport energy – and, if they're locally traditional, are proven suitable for the local climate.

Embodied energy in buildings (typical)[5]		
Steel buildings	300,000 BTU/ft^2	8.17 kWh/m^2
Concrete buildings	200,000 BTU/ft^2	5.45 kWh/m^2
Wooden buildings	40,000 BTU/ft^2	1.09 kWh/m^2

A steel building typically embodies 7½ times the *energy* of a wooden building, but embodied *carbon* is infinitely higher as all-wood buildings are carbon negative.

Common building materials: embodied impacts

Material	Relative embodied energy per weight[6]	Other impacts from production	Comments
Timber	1	Depending on forestry practice, impacts range from the very negative (e.g. soil acidification, and denutrification, landslides) to the positive (e.g. habitat improvement, flood prevention).	Locks up carbon for its lifetime.
Bricks	4	Clay pits; energy intensive; smoke includes sulphur oxides, nitrogen dioxide, carbon monoxide, CO_2, and black carbon particulates[5].	Long, maintenance-free life.
Cement	5	Very fine airborne particulate which travels huge distances.	CO_2 production ranges from moderate to very high – depending on production process.
Plastic[8]	6	Highly toxic pollution.	Prone to UV deterioration, but non-biodegradable; thermoplastics are recyclable (in principle, not always in practice).
Glass	14	Water pollution, NO_x, SO_x and dust air-pollution.	Non-biodegradable but widely recycled.
Steel	24	Sulphur dioxide, nitrogen dioxide, carbon monoxide, soot and dust.[9]	Widely recycled (but this causes toxic pollution from coatings).
Aluminium	126	Fluoride (highly toxic), CO_2 and perfluoro-carbons: potent, long-lived greenhouse agents.	Very widely recycled due to expense.

Keypoints

- (New) buildings' embodied energy – and hence embodied-CO_2 – now averages a third of their lifetime operating energy.
- Embodied-energy's climate impact is effectively double that of comparable operating energy.
- Natural materials' embodied energy/pollution is lower than manufactured materials'.
- Local natural materials embody little transport energy.
- Use of reclaimed, recycled and waste materials for building further reduces embodied energy/pollution.

Choices

Options	Decision-making factors
Local or distant materials?	Embodied energy; suitability; local identity; availability; transport cost and CO_2.
Natural or manufactured materials?	Embodied energy; performance; durability; health effects; soul-nourishment; cost.
New, second-hand or recycled materials?	Embodied-energy; availability; quality; design flexibility; labour cost; product cost.

Things we use

Energy-wise, the things we use *in* buildings are surprisingly significant. In super-insulated homes, most energy consumption is due to consumer electronics and white goods.[10] Undemanding lifestyle-changes (e.g. walking instead of driving to shops, using energy-efficient appliances and lighting, turning off televisions and similar 'vampire loads' instead of leaving them on standby,[11] running the washing machine when sunshine is producing photovoltaic electricity, and air-drying laundry) actually save slightly *more* energy than insulating buildings.[12] Consequently, if you can't make changes to the building itself (for instance, if you rent your home) you can still do lots to reduce your environmental impact.

Despite their slightly higher purchase cost, the lower energy consumption of Grade A appliances (and, for washing machines, water-saving) saves money over their lifetime. Low-energy lighting delivers about 50% better CO_2-reduction/£ than building insulation.[15] I much prefer LED lamps to compact fluorescents (CFL). All fluorescents contain toxic mercury vapour. Although the quantity is small, dud bulbs and tubes should be treated as toxic waste – but most people

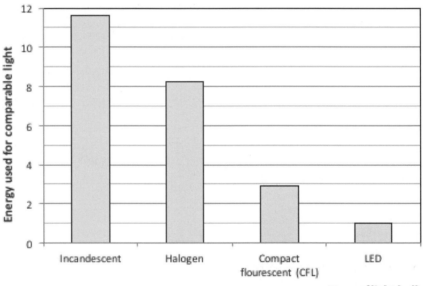

Type of light bulb

Light bulbs: efficiency/watt[13]

Type	Normal range (Lumens per watt)	Maximum (Lumens per watt)	Comments
Incandescent	12.5-17.5		Manufacture now banned in EU.
Halogen_	16-24		2,000-hour lifespan.
Compact fluorescent (CFL)	45-75	100	6,000–15,000-hour lifespan. Health concerns (sub-sensible flicker, mercury vapour).
LED	30-90	200	25,000-hour lifespan. (Currently) expensive but 90% power-saving means short payback.[14]

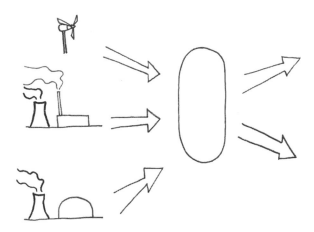

Green electricity? Or greenER electricity?

Keypoints

- Consumer electronics and white goods account for most of super-insulated homes' energy consumption.
- Undemanding lifestyle-changes and use of energy-efficient lights and appliances save slightly more energy than insulating buildings.
- Low-energy lighting reduces 50% more CO_2/£ than building insulation.
- Grade A appliances save money over their lifetime.
- Using alternatives to electric devices reduces CO_2-production.
- Appliance longevity is also significant.
- Sharing facilities and equipment produce the biggest energy – and money – savings.

Choices

Options	Decision-making factors
Lifestyle-changes, energy-efficient lights and appliances or building insulation?	Convenience; time-saving; first cost; lifetime cost; thermal comfort.
Grade A or cheaper appliances?	Purchase cost versus lifetime cost; longevity; reliability.
Low-energy or cheaper lighting?	Purchase cost versus lifetime cost; longevity; lighting quality, colour, mood.

just put them in the bin. If you break one indoors, it's serious. Leave the room fast, shut the door, take a deep breath and, without breathing, return to open the windows.[16]

Appliance longevity is also important: manufactured goods embody about half the energy they use over their lifetime,[17] so the longer they last, the less CO_2 produced by replacing them. As heat shortens light bulbs' life, lifespan reflects energy efficiency: doubly recouping their increased purchase cost.

All such appliances use electricity. Electricity dependence is steadily growing. We now own, on average, three and a half times as many electric appliances than in the 1970s.[18] Naturally, eco-homeowners buy 'green electricity', but how green is this? Distribution grids mix all sources: renewable, fossil-fuelled and nuclear. The more consumers buy green electricity, the more in the mix, but it's still a mix: mostly fossil-fuel- or nuclear-generated.[19] Not using electricity, or using non-electric devices instead of electric- or petroleum-powered ones is a big step towards CO_2 reduction – and a big money-saver too.

The biggest energy and money savings, however, come from sharing facilities (see: **Chapter 4: Choosing to group homes**, page 35). Car clubs are another form of sharing, increasingly popular. These save members an average of £3,500/year (in the UK; in USA, $9,000/year) – besides saving the CO_2 embodied in the cars not needing to be bought.[20]

Things we get rid of

Buildings are only shells through which life, energy and things flow. Life imprints its record. Energy degrades into heat and eventually dissipates into outer space. But what about things? They remain on Earth. Where do they go? What happens to them? Can they be re-integrated into natural cycles?

Water recycling

The most obvious thing that flows through buildings is water. Water is cheap – but it's far from valueless. It's indispensable for life: certainly not something to waste. We assume it's plentiful but in many areas, there soon won't be enough.

Water conservation starts, of course, with repairing leaks. Plumbing layout is also important: short hot-water piping dead-legs minimise the cooled water that must be run off before hot comes through – saving both energy and water. Appliances too can save lots of water. Whereas pre-2001 British and pre-1997 American toilets use 13-litre flushes and modern ones 6 litres, some symphonic ones (e.g. ES4, Ifö) use only 4: 33-70% savings. Dual-flush retrofit kits for cisterns are available, so only faeces need get a heavy flush. For taps, flow-restrictors reduce flow-rate; spray heads (and mister showers) entrain air, so less water is needed. For any spray and mist devices, heated water needs briefly heating to 60°C (140°F)[21] to kill Legionella bacteria before mixing with cold for non-scald tap temperature (for showers, 41°C/106°F; baths, 44°C/111°F[22]).

For irrigation, rainwater is much better than tap water – and free! For this, simple rainwater butts suffice for both collection and storage. For indoor use, only clean roofwater is suitable. Of this, the dirt-rich first-flow must be discarded. (Living-roof water, although filtered, is brown, so unacceptable indoors.) To avoid cross-contamination and misuse, rainwater needs a wholly separate piping system, colour-coded with labelled taps. Being 'soft' (mildly acidic, so better at dissolving soap), it's better for laundry than most mains water. It also suits other non-potable uses. Sterilised, it's drinkable but, like distilled water, lacks trace-minerals so is inadvisable for long-term consumption.[23]

There are well-proven rainwater-collection systems manufactured; and inexpensive DIY systems aren't hard to improvise. As the plastic tanks in (all-new) single-house systems account for much more CO_2 than ordinary mains water supply, however, rainwater-harvesting's benefits are disputed.[24] *Reused* plastic, or concrete, tanks – and/or communal systems, however, minimise carbon-footprint.

The usual size of tank for a (British) single house is 1,500–2,000 litres, but you can't store all the rainwater you collect as, to keep it fresh, tanks must flush often.

They're better, therefore, too small than too large.[25] If there's not enough rain, they can be topped-up with mains water, using a motorised supply valve controlled with a float switch.[26]

Terrace houses with roof-ridges parallel to the street can only easily collect rain from half the roof.[27] Normally, however, we discard it all. Consequently, virtually all rainwater runs off roads and roofs instantly, often causing floods. Permeable paving, gravel and (some types of) brick paving let rain soak into the ground. Green roofs spread run-off over 24 hours. Green roofs, brown roofs and most kinds of green (living) wall, however, rely on waterproof membranes – which have high embodied pollution (some more than others; see: **Chapter 10: Insulating roofs**, page 108). Nonetheless, green roofs' flood-control benefit often outweighs this adverse environmental impact. For single houses, green roofs' stormwater run-off is slow enough to irrigate gardens without washing soil away. None needs to go into sewers. This reduces water rates: so saves money (albeit unfairly little) every year. In larger projects, savings on stormwater solutions and energy reductions pay back costs more quickly.[28]

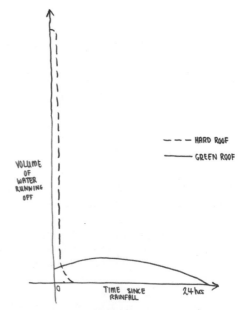

Impermeable roofs and paving: instant run-off, causing floods. Green roofs spread run-off over 24 hours, reducing flood risk.

Greywater is everything except sewage that goes down the drain: some two-thirds of all household water.[29] It can't be stored as it would putrefy, but it can be used to irrigate gardens. Although its pathogen content is normally low, faecal pollution *is* possible (e.g. from cleaning babies, soiled laundry). Irrigation, therefore, should be sub-surface and limited to non-edible plants: preferably crops grown solely for composting (e.g. nettle, comfrey, vetch, alfalfa). Distribute it through rainwater land-drains and soak-aways, so that rainwater dilutes alkalinity build-up in the soil. To further minimise alkalinity, use eco-friendly washing powders and washing-up liquids instead of detergents. To ensure non-contamination by sewage, greywater needs a totally separate plumbing system.

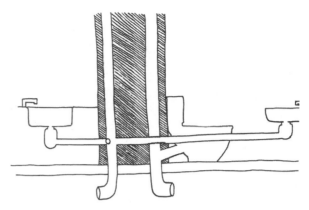

Greywater waste-plumbing separate from sewage.

Nutrient recycling

The other third of domestic water ends up as sewage. This would decompose naturally, were it, and human populations, not so concentrated. Composting toilets and reedbeds emulate natural decomposition. Composting toilets were originally developed to save water.[30] They mix pathogen-rich faeces with kitchen waste (safe for garden compost) and urine (which starts out pathogen-free) so, if breakdown isn't complete, the finished compost *isn't* safe to put on edible crops or anywhere children might play. This wastes the safe ingredients – which could otherwise be used as nutrients. If, however, you can collect uncontaminated urine (e.g. by using separating toilets), it's usable as a compost-accelerator or (diluted 1:10-20) for fertiliser. This has been done at apartment-block scale in Sweden since the 1980s, so is proven safe. With separating toilets, men must take trousers down and sit to urinate into the right part. As some don't bother, it's best to add a urinal in the toilet room. Urinals can be air-flushed using infinitesimal amounts of energy. With fan-driven extract ventilation, composting toilets shouldn't smell any more than normal ones (against which, instead of spraying chemicals, strike a match to ionise the air).

Alternatively, you can use conventional toilets but decompose the sewage in reedbeds. Reedbeds also don't smell. They don't need much attention but do

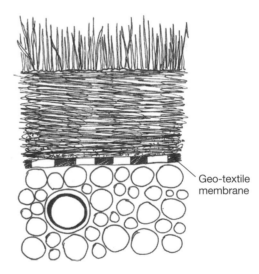

Geo-textile membrane

Sub-surface irrigation using land-drain pipes.

need space: a settlement tank at least 15m (16½ yards) from the house, then at least 1m² (1⅕ square yards) / person of reedbed, an effluent tank, then 100m² (120 square yards) of willow bed – i.e. in total, at least 35m (38 yards) to your boundary.[31] Except on rural sites, however, composting toilets, reedbeds or other sewage treatment methods rarely justify their space, expense, complications and management.

Food and garden waste is different. It's easy to compost this, in either heaps or bins. Instead of wasting the nutrients by mixing them with sewage, this captures them for use in the garden. Just as important, composting also avoids putrification in landfill,

With two-chambered compost bins you can fill one; when this is full, turn the contents into the other, from which you remove finished compost when it's ready. Compost bins are easy to make (e.g. from scrap pallets). With long heaps, fill at one end, remove from the other.

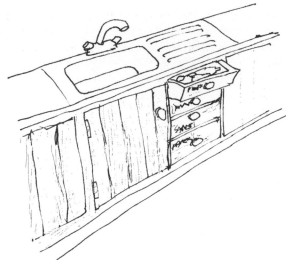

For non-compostable waste, under-sink cluster-bins or drawers ease collection for recycling.

which produces methane, a gas 21 times more climate damaging than CO_2.

Although compact forms are warmer, which accelerates decomposition, long compost heaps let you add material to one end and remove finished compost from the other. Even better are parallel heaps, so you can turn the compost so that the outside (where weed seeds sprout) becomes inside (where they decompose). Similarly, with two-chambered bins, you can turn the contents from the first to the second, and let it mature while you refill the first chamber. Heaps or bins, however, shouldn't get *too* hot: this destroys nutrients, kills micro-life and risks spontaneous combustion.

For both moisture and drainage, and to build up a resident worm and micro-life population, the contents need good contact with the ground. Shade is desirable but tree shade isn't: tree roots steal most of the nutrition! If you have a rat-proof container (which chicken wire isn't), you can put any kind of ex-living stuff in. If not, meat and cooked food attracts vermin (e.g. rats, seagulls) and neighbours' dogs, so exclude these. Optimally, there should be similar amounts of nitrogenous matter (e.g. weeds, kitchen peelings) and

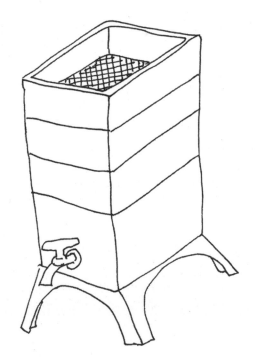

Worm composting.

carbon-rich matter (e.g. cardboard, woody plant stems). Too much carbon (particularly paper,[32] twigs, sawdust, straw) drastically slows the composting process. Nitrogen-rich constituents (e.g. coffee grounds, legumes, manure, urine) speed composting. So does lime (e.g. left over from building) – but don't use it *with* manure as this destroys manure nutrients.

If you have a large kitchen or warm basement, you can even compost indoors, using worms. This is ultra-compact and ultra-fast: it only takes a month. Basically, wormeries comprise a stack of perforated trays over a bowl with a tap to drain the liquid produced for use. This liquid is strong, so use sparingly or dilute with water. When you've filled one tray with vegetable waste, put another on top; the worms come up through the perforations. When you've filled all the trays, the lowermost one is ready to use as compost.[33]

Keypoints

- Keep greywater waste plumbing separate from sewage, so greywater can be used for irrigation.
- Consider urine collection for fertiliser.
- Harvest rainwater:
  - Collect it for irrigation, and possibly other non-potable uses.
  - For indoor use, only collect clean roofwater.
  - Colour-code piping and taps.
  - Size tanks for frequent overflow.
- Use recycled tanks if possible.
- Buildings can cause floods. In flood-prone watersheds, use permeable paving and consider green roofs.
- Compost food/garden waste.
- Use second-hand, recycled and waste materials for building.

Choices

Options	Decision-making factors
Individual or communal compost-making?	Private or communal gardens/plots; garden size; proximity to communal composter / collection-point; volume; allocation of finished compost.
Heap, bin or worm composting?	Space to turn heaps and for bins; warm indoor space for worms; attention demands (e.g. warmth/watering during holidays); speed.
Separated greywater and/or urine or mixed sewage?	Feasibility of separation; urine storage or direct use (e.g. as compost accelerator); greywater alkalinity; water availability/price.
Rainwater harvested, directed to irrigation or absorbed by green roofs?	Rainwater need; for what use; rainwater volume; soil/drainage characteristics.

Construction waste reuse

Another high-volume waste-stream is construction waste: a quarter of all (UK) waste, 100 million tons/year.[34] This has no nutrient value but much of it is far from useless. Some waste from building and renovation work (e.g. paint strippings) is toxic so must be disposed of safely, but some can be sold (e.g. iron fireplaces and baths, lead and copper); some can fuel wood stoves (e.g. damaged timber, laths); and much can be reused on-site (e.g. demolition rubble for hardcore, excavation earth for ground modelling). The greatest recycling benefit, of course, comes from using waste (e.g. timber off cuts) and second-hand materials (e.g. reclaimed tiles, bricks, timber, doors, sinks, etc.) for building. This may sometimes require you to modify your design to accommodate what is available. Likewise, living a waste-reusing life unavoidably modifies lifestyle, but this needn't be onerous or unduly time-consuming, and it does save money, not to mention the planet.

Things we affect: biodiversity

Biodiversity depends on biodiverse habitat. Hence 'wild', unkempt bits of gardens attract more animal wildlife (from butterflies and other insects to birds and mammals) than manicured parts do. Do insects really matter, though? For songbirds, they're food. For all of us, pollinators ensure our food supply. Worldwide, honeybees (the best of all pollinators) are in decline – with potentially catastrophic consequences. Indeed, 85% of British people consider this more alarming than climate change.[35] As agricultural pesticides decimate bee populations, bee-friendly gardens are increasingly important. Although it may not seem possible to keep bees safely in towns, they fly in 'bee lines', namely straight, so fences, dense hedges or tube outlets from hives can keep them safely above head-level. Beekeeping, however, requires equipment, indoor space and commitment. It's easier just to plant a bee-friendly garden. For this, the Royal Horticultural Society publishes a list of suitable plants.[36]

All buildings replace natural habitat with something man-made. Ecologically, this may not matter much in the countryside (although its *visual* impact might), but in towns, the aggregate effect is major. Buildings, however, can also *create,* or even improve, habitat for some creatures, particularly songbirds and bats.

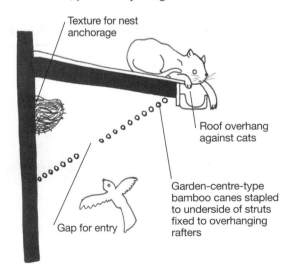

Texture for nest anchorage

Roof overhang against cats

Garden-centre-type bamboo canes stapled to underside of struts fixed to overhanging rafters

Gap for entry

Bamboo-closed soffit for eaves-nesting birds.

Despite their Transylvanian associations, bats fulfil an essential ecological role – which is why they're protected species. Besides eating massive amounts of insects (especially important for us is their consumption – and deterrence – of mosquitoes and crop pests), they're also pollinators.[37] Old buildings let bats access warm spaces through cracks.[38] New ones need deliberate 20mm (³/₄") apertures (e.g. board-gaps, bat-bricks, ventilation tiles). Felted-fibre breather membranes, however, can snare and kill bats.[39] Either mesh-cover openings to deny them access or, for established nesting sites, use traditional bituminous underfelting and good ventilation.[40]

For songbirds, green and brown roofs provide homes for the insects off which they feed. Conversely, replacing lawns with decking and gravel reduces their forage opportunities. Traditionally constructed buildings have rough surfaces for birds' claws to grip whereas modernistic glass, concrete and steel are too smooth.

Birds need nesting and roosting places (often only a few metres up), food, water and safety.[41] Besides hedges, trees and nest boxes, birds nest in climbers, under-roof spaces and in small apertures in walls (from damaged masonry to specifically formed 'swift-bricks'[42]), especially where sheltered from hot sun and cold winds. Evergreen climbers also provide sheltered roosts in winter. If these climbers cloak around a building's corners, birds only need to shift slightly around the corner when the wind direction changes.[43] Crevices and overhead-shields (e.g. roof-eaves) protect nests; and lookout perches (e.g. roofs, trees, fences) let them see when it's safe to feed or drink.[44] Eaves-nesting birds, like swallows, appreciate anchor-points (e.g. nails, metal-lathing, rough textures).[45]

To enjoy watching birdlife, locate nest-supports, nest boxes and bird-feeders (which should be inaccessible to squirrels) in view from windows. Birds also require lookout perches near feeders and birdbaths: often they use laundry lines for this, leaving birdshit on the laundry in the process. To avoid this, provide lookout perches that give them a better view than your laundry line does. Similarly, for birdshit-free people, ensure routes between lookout perches and nests and feeders aren't above doorways or paths.

Aquatic plants for ponds[46]

Plant group	Ecological role		Importance
Submerged	Oxygenate and filter.		Essential
Floating-leaf[47]	Filter and shade.	Cool water has greater oxygen-retention capacity.	Essential
Floating			Preferable
Marginal	Filter through their roots.		Preferable

For bird safety, joined-up greenery minimises exposure to hawks, and impenetrably thorny shrubs minimize danger from cats. Multi-level canopy, maximum 'edge' conditions and dead wood maximise forage opportunities. If on or near the ground, foraging, drinking and bathing sites need sufficient open ground around them to avoid ambush by cats. Birds' bathing sites (e.g. birdbaths, ponds) also need sunlight so feathers can dry.

All living things need water in one form or another. Not surprisingly, therefore, ponds and streams increase an area's biodiversity. These don't have to be on your land but if they are, you'll see more wildlife. Also, the more pleated their edges, the greater the habitat variety: hence more species. The larger they are, the less likelihood of drying up or freezing solid.

Water, of course, is never problem-free. For child-safety it must be very shallow (100mm / 4")[48] protected by a grill or inaccessible to small children. (But children are ingenious; is *guaranteed* child-inaccessibility ever possible?) Additionally, water plus sunlight and nutrient means algae. Zooplankton, snails, tadpoles and many fish species eat algae. Water-purifying plants consume nutrient faster than algae, limiting algal growth. Shading (e.g. by floating-leaf plants), aeration and, for deep water, turbulence also help. There are also ultra-violet sterilisers, but this is the electrical (CO_2-producing) way.

If water is still for a week or more, you'll get mosquitoes. Mosquitoes can't lay eggs in moving water, although the amount of still water they need is very small; for instance, water pooled in leaves next to a stream is enough. In ponds, you can introduce fish that eat their larvae, but still water needs to be kept clean and alive. Besides protecting it from nutrient contamination (e.g. fallen leaves), to remain healthy, it needs aquatic plants.

Keypoints

- Plant a bee-friendly garden; and maximise (bird-food) insect habitat.
- Birds need safe nesting, roosting, feeding and drinking/bathing opportunities.
- Multi-level canopy, maximum 'edge' conditions and dead wood maximise songbirds' forage opportunities.
- Crevices, thorny shrubs and overhead-shields protect nests; and lookout perches let birds see when it's safe to feed or drink.
- Traditionally constructed buildings have rough surfaces for bird-claw grip and fissures for (bird-food) insects; modernistic glass, concrete and steel don't.
- Provide anchor-points for eaves-nesting birds.
- Locate anchor-points, nest boxes and feeders in view from windows but not above doorways, paths or near laundry lines.
- Bats need access to warm spaces through cracks, board-gaps, bat-bricks, ventilation tiles, etc.
- Felted-fibre breather membranes snare bats, which can kill them.
- For small-child safety, keep water very shallow (100mm / 4") or inaccessible.
- Protect ponds against algae by nutrient exclusion, aquatic plants and fauna, aeration and/or using ultra-violet sterilisers.
- Protect against mosquitoes by denying suitable habitat (e.g. no still water), stocking ponds with mosquito-lavae-eating fish – and encouraging bats.

Choices

Options	Decision-making factors
Bird and bat access to eaves-soffits, roof voids, sheds?	Tolerance to droppings; bird, egg and bat safety (e.g. from cat ambush, bigger birds, rats, claw-snaring felts).
Where to locate nest boxes, nest-anchorage points and feeders?	Everyday view; droppings avoidance; bird safety from cats, feeder- and nest-robbers.

Resources

Wildlife habitat

Aquatic plants/ecology
Á fleur d'eau (catalogue)

Bee habitat
www.rhs.org.uk/advice/pdfs/plants-for-bees
www.gardenersworld.com/plants/features/wildlife/
plants-for-bees/1107.html

Bat habitat:
arbtech.co.uk/bats-breathable-roof-membranes-
need-know
www.batsandbrms.co.uk/updates.php
www.bats.org.uk/pages/bat_roosts.html

Bird habitat:
Barrington, R. (1971) *The Bird Gardener's Book*,
Wolfe Publishing, London.
Johnston, J. and Newton, J. *Building Green,* London
Ecology Unit.
www.rspb.org.uk/discoverandenjoynature/
discoverandlearn/birdguide/name/s/swallow/
encouraging.aspx
www.swift-conservation.org/Nestboxes&Attraction.
htm

Material impacts
Information
Berge, B. (2nd ed. 2009) *The Ecology of Building
Materials*, Architectural Press, Oxford.
Day, C. (3rd ed., 2014) *Places of the Soul*,
Routledge, London

Day, C. (2002) *Spirit & Place*, Architectural Press,
Oxford.
Desai, P. (2010) *One Planet Communities,* John Wiley
& Sons, Hoboken, NJ.
Energy Saving Trust, Powering the nation report
CO332(1).pdf.
Griffiths, N., (2012) *Eco-house Manual*, Haynes
Publishing, Yeovil.
Harland, E. (1993) *Eco-Renovation*, Green Books,
Devon.
Car club*: www.zipcar.com*
Lighting: *www.rapidtables.com/calc/light/how-
lumen-to-watt.htm*

Recycling/reuse
The Greenshop, Rainwater Harvesting.
Thompson, K. (2nd ed. 2011) *Compost*, Dorling
Kindersley, London.
Wack, H.O., Rainwater utilization in housing, Wisy.

Sewage treatment
www.clivusmultrum.co.uk
www.doeni.gov.uk/niea/ppg04.pdf
www.soilassociation.org/LinkClick.aspx?fileticket...
tabid=151
www.sustainablebuild.co.uk/composttoilets.html

Suppliers of less common green products
www.backtoearth.co.uk
www.eco-building-products.co.uk
www.ecologicalbuildingsystems.com
www.ecomerchant.co.uk
www.greenbuildingstore.co.uk
www.greenbuildingstore.co.uk/page-airflush-
waterless-urinal.html
www.greenshop.co.uk
www.natural-building.co.uk

Less common products
Heat recovery:
www.renewability.com/power_pipe/index.html
Rainwater harvesting: *www.rainharvesting.co.uk/
products/filters?view=featured*
Vermiculture: http://www.originalorganics.co.uk/
wormeries

1. And other greenhouse gases.
2. Somebody has laboriously worked this out at 47%. But add the transport buildings generate and it approaches three-quarters.
3. 30-50%: Berge, B. (2000) Some ecological aspects of building materials, in Roaf, S.; Sala, M. and Barstow, A. (eds) TIA Conference 2000 proceedings. It used to be an eighth, but modern buildings are more energy efficient, are built of more-processed materials and have shorter lifespans, 60 years on average: Cotterell, J. and Dadeby, A. (2012) *PassivHaus Handbook*, Green Books, Cambridge.
4. de Selincourt, K. (2012) Embodied energy – a ticking time bomb?, *Green Building* magazine, Spring 2012, Vol. 21, No. 4.
5. McDonough, W.,(1993) A boat for Thoreau: Architecture, ethics and the making of things, *Business Ethics*, May/June 1993.
6. Assuming virgin new material: Harland, E. (1993) *Eco-Renovation*, Green Books, Devon.
7. *http://scienceblogs.com/thepumphandle/2013/08/21/building-a-better-brick-kiln-to-fight-pollution* (accessed: 10 December 2013).
8. Embodied energy in plastics, however, varies greatly: relatively low in 'simple' ones (e.g. polythene, polystyrene, polypropylene) but high in more complex ones (e.g. polyisocyanate, polyurethane): Broome, J. (2007) *The Green Self-build Book*, Green Books, Devon.
9. *www.ilo.org/iloenc/part-xi/iron-and-steel/item/593-environmental-and-public-health-issues* (accessed: 10 December 2013).
10. Bothwell, K. (2013) Housing retrofit, *Green Building* magazine, Spring 2013, Vol. 22, No. 4; Smith, R. (2007) Eco-renovation, *Green Building* magazine, Winter 2007, Vol. 17, No. 3. TV-watching takes 10% of Californian electricity, BBC World Service, 19 November 2009.
11. This halves such appliances' power consumption: MacKay, D. (2009) *Sustainable Energy – without the hot air*, UIT, Cambridge.
12. Insulation: 4-9 megatonnes of CO2/year; lifestyle and energy-efficient appliances, 5-9: Welsh Assembly Government (2009) Climate change strategy – programme of action consultation.
13. *www.rapidtables.com/calc/light/how-lumen-to-watt.htm* (accessed: 10 November 2013)
14. Down to one year for frequently used lamps.
15. Griffiths, N. (2012) *Eco-House Manual*, Haynes Publishing, Yeovil.
16. And air the room for at least 10 minutes; don't use a vacuum cleaner: *www2.epa.gov/cfl/cleaning-broken-cfl* (accessed: 8 September 2014).
17. 30-50%: Berge, Some ecological aspects.
18. Energy Saving Trust, Powering the nation report CO332(1).pdf (accessed: 11 January 2014).
19. Globally, 67% is from fossil fuels, 12% nuclear: Triodos Bank (2013), The Colour of Money, Autumn 2013.
20. For 6000 miles/year: Desai, P. (2010) *One Planet Communities*, John Wiley & Sons, Hoboken, NJ; *www.zipcar.com*. Some hundred American cities have these.
21. Two minutes at 60°C (140°F) kill 90%; 70°C (158°F) kills all instantly. Worcester boilers, however, recommend 65°C (150°F) or UV sterilisation.
22. *www.ciphe.org.uk/Global/Databyte/Safe%20Hot%20Water.pdf* (accessed: 22 December 2013).
23. Other health concerns are distorted molecular memory (Dr Masaru Emoto: *www.hadousa.com*) and metal-dissolving acidification by dissolved CO_2 (US Environmental Protection Agency).
24. Thornton, J. (2008) Rainwater harvesting systems, *Green Building* magazine, Spring 2008, Vol. 17, No. 4, G
25. Tanks are better too small (with minimal mains-water replenishment) than too large (UK rule of thumb: $1m^3$/25-40m^2 roof or 18 days' need): The Greenshop, Rainwater Harvesting; Simpson, J. (1998) Stopping water going down the drain, *Building Design*, 13 March 1998.
26. When the tank is full, the flotation forces are too great for a normal ballcock.
27. You can, of course, pipe it from one side to the other, but as the piping is cold, so must be insulated, this takes a lot of space.
28. *www.environment-agency.gov.uk/business/sectors/91970.aspx* (accessed: 16 December 2013). To minimize flooding, Germany has tax incentives for green roofs.
29. (Or half): Using graywater for landscape irrigation, *Environmental Building News*, March/April 1995, Vol. 4, No. 2.
30. For holiday homes on Swedish islands: mostly rock, so little soil-life to decompose faeces, and little water to spare for flushing sewage into the sea that holiday homeowners enjoy swimming in.
31. *www.doeni.gov.uk/niea/ppg04.pdf* (accessed: 26 December 2013).
32. It's worth noting that glossy paper includes a lot of (inert, so non-nutrient) clay, and many yellow and red printing inks contain heavy metals.
33. *www.originalorganics.co.uk/wormeries* (accessed: 14 January 2014)
34. Stow, G, McTiernan, D. (2008) Space for waste, *Green Building* magazine, Spring 2008, Vol. 17, No. 4.
35. The British honeybee population has halved in 25 years: Triodos Bank (2014) *The Colour of Money,* Summer 2014.
36. *www.rhs.org.uk/advice/pdfs/plants-for-bees*; see also: *www.gardenersworld.com/plants/features/wildlife/plants-for-bees/1107.html*
37. *www.blm.gov/id/st/en/environmental_education/BLM-Idaho_nature/wildlife/bats/the_benefits_of_bats.html*

38. Preferably east-facing walls as solar–warmth takes about four hours to pass through bricks: Johnston, J. and Newton, J. *Building Green*, London Ecology Unit.

39. Breathable membranes' fibres can fatally entangle bats: Waring, S. (2010) Breathable roofing membranes, *Green Building* magazine, Summer 2010, Vol. 20, No. 1.

40. *arbtech.co.uk/bats-breathable-roof-membranes-need-know* (accessed: 16 June 2014). Currently, only BITUMEN TYPE 1F felt is safe for bats: *www.batsandbrms.co.uk/updates.php* (accessed: 16 June 2014).

41. Barrington, R. (1971) *The Bird Gardener's Book*, Wolfe Publishing, London.

42. Swift slots: 65mm wide x 30 high; swallow, minimum 200 x 50; bat, 20–50 x 20 (or 45-diameter hole): *www.swift-conservation.org/Nestboxes&Attraction.htm*; *www.rspb.org.uk/discoverandenjoynature/discoverandlearn/birdguide/name/s/swallow/encouraging.aspx*; *www.bats.org.uk/pages/bat_roosts.html* (accessed: 14 April 2013).

43. Barrington, *Bird Gardener's Book*.

44. Ibid.

45. Not over doors or sitting places!

46. Á fleur d'eau (catalogue), Edition 96.

47. Lilies need sun five hours/day to flower, so shouldn't be under trees: ibid.

48. Olds, A. R. (2001) *Child Care Design Guide*, McGraw Hill, New York.

Generating energy

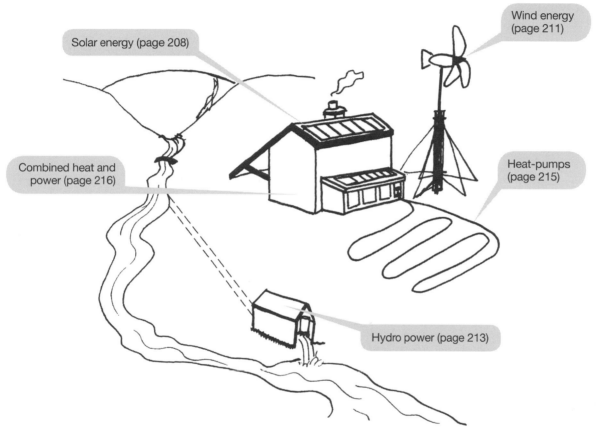

CHAPTER EIGHTEEN

Solar energy (page 208)

Wind energy (page 211)

Combined heat and power (page 216)

Heat-pumps (page 215)

Hydro power (page 213)

Micro-energy generation basics

In crude terms, it's energy production that is destroying our climate. So can we produce it ourselves from renewable resources? Yes. But it's always cheaper, more effective – and usually much easier – to conserve energy than produce it. So only add energy generating devices after you've done everything you reasonably can to prevent heat loss and unwanted heat gain, and to cut down on electricity usage. Most domestic-scaled energy generating technologies produce heat *or* electricity; only two technologies produce both. The most common ones are:

- For heat: solar heating (both of space and water) and heat pumps.
- For electricity: photovoltaics (PV), wind-power, and hydro power.
- For both electricity and heat: combined heat and power (CHP), and some PV panels.

Seasonally-matched systems[1] for a typical non-tropical climate			
Hydro, Solar	Solar	Wind	Hydro, Wind
Spring	Summer	Autumn	Winter

Solar power and heat is available almost everywhere, although not always commercially feasible. Wind- and hydro-generators' suitability is largely limited to rural sites: in towns, rooftop wind is weak and turbulent. Heat pumps use electricity, so, if they're to reduce carbon emissions, they need to be high-efficiency and only required to boost temperature a little – and/or use renewable-powered electricity. CHP needs a heat source (usually fuel, but possibly solar): at a small scale, it's only economical for off-grid homes.

Sun, wind and (sometimes) water are weather and/or season dependent, so each season favours one kind of energy. Solar and wind energy are intermittent, so need to be stored. Electricity can be sold to mains electricity suppliers when you're generating more than you're using, then (more expensively) bought back when you need it. Alternatively, the electricity generated can be (even more expensively) stored in batteries or (prohibitively expensively at single-house scale) used to pump water uphill to supply a hydro-generator; or used to heat water. Solar heat can be stored in thermal mass. This has implications for design and retrofit (see: **Chapter 8: Thermal mass**, page 83; **Solar heating**, page 86).

Keypoints

- It's always cheaper to conserve energy than to produce it.
- Only add energy generating devices after all feasible energy-conservation measures.
- For wind generators, urban rooftop wind is too weak and turbulent.
- Solar power and heat is available almost everywhere.
- Sun, wind, and (sometimes) water, is weather- and/or season-dependent, so needs to be stored (e.g. by mains feedback for electricity; thermal mass for heat).

Solar energy

The sun is the source of all energy on earth, both legacy (geo-thermal, fossil fuel) and current (sunlight, wind, biomass, etc.). Sunlight spans a broad spectrum: 52-55% is infrared heat; 42-43% is visible light; and 3-5% is ultraviolet. For heating water or home interiors, we can use the heat wavelengths (see: **Solar heating**, page 86). But it's *light* that produces photovoltaic electricity. Heat, in fact, compromises PV efficiency.

As direct sunlight is radiant, it's partially blocked and scattered by clouds, which are semi-translucent; but wholly shaded out by opaque solid things, leaving only diffuse daylight. For both heating and photovoltaic technologies, therefore, (largely) unshaded locations are crucial. Consequently, the higher panels are on roofs, the better. Most water-heating panels must be at least 60mm (2½") above tanks, as they drain down to avoid frost damage. For (less efficient, but wholly non-electric) thermo-syphon systems, however, this is reversed: the tank must be at least 900mm above collector panels. (Pumped systems can be photovoltaic powered, so don't need mains electricity.[2]) Besides shade avoidance, you also need to consider ease of access for scaffolding erection (for installation), potential tree growth (hence shade) – and whether this can be managed by pruning (e.g. are trees yours? Can you get to them? Where will pruned-off branches fall? Will you be safe?).

Manufactured solar water and photovoltaic panels are rectangular, so rectangular roof planes (i.e. mono- or bi-pitched roofs) can fit more panels on them than can hipped, odd-shaped or curved roofs.[3] Besides solar panels, though, lots of other things need sunlight. Gardens do for plant growth. We do for enjoyment. Our homes (through windows) do for warmth and mood. This makes shade an issue. Hipped roofs cast less than simple bi-pitched roofs, particularly when sun is low. You need, therefore, to consider the shading consequences of roof design as well as its suitability for solar panels.

Solar panels should face the sun. Photovoltaic panels can be on motorised tracking mounts, but this is expensive. Cheaper, and only a little less efficient, is to fix their orientation and manually adjust inclination two (or more) times a year. Freestanding arrays, however, shade the ground – which needs sunlight for crops or amenity. Most PV, and all water-heating panels, therefore, are roof-mounted. Optimally, all kinds face due south.[4] But what if roofs are diagonally aligned? East orientation only decreases efficiency 15%,[5] but this matters more for PV than for water heating, as PV panels are more expensive. For PV, south-east orientation is preferable to south-west as it's cooler – increasing efficiency[6] – and often sunnier as convection clouds form later in the day. (This, however, depends on locality: some places have morning mists.) For solar-water panels, south-west orientations are preferable to south-east: warmer and better timed, as more hot water is used in evenings or early mornings than at noon.

Optimum inclination also differs, although isn't critical. More hot water than you need is no use. Consequently, solar water-heating panels are best inclined for the period when there's *only just enough* hot water, namely incident to equinoxial noon sun:[7] in Britain, around 42°. For photovoltaics, if electricity can be fed back to the grid, annual yield is your focus. As yield is some five times higher in summer than winter, it makes sense to incline panels for summer collection (i.e. a little shallower).[8]

For both kinds of panel, there's a confusingly extensive range of efficiencies and prices. For photovoltaics, unless roof area is unduly limited, you usually get better value for money by using a larger area of less expensive, less efficient ones. Additionally, above-roof PV arrays allow airflow under them so keep cooler than in-roof ones. This improves output some 3%, but at the price of aesthetics. Although photovoltaic output is normally limited by the lowest-performing (ie most shaded) panel,[9] multi-string inverters or 'optimisers' can overcome this.[10]

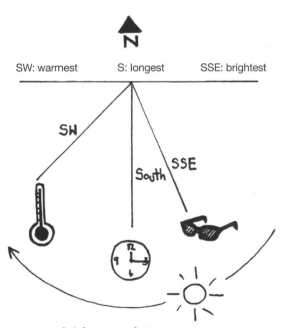

Longest sun, brightest sun, hottest summer sun.

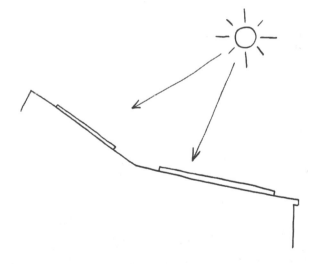

Optimally, solar water-heating panels are best inclined for the period when they can only produce just enough hot water. But photovoltaic yields are much higher in summer than at equinox, so, if electricity is fed back to the grid, incline PV panels for summer collection. Usually, however, panel inclination is fixed by roof pitch – which is determined by other factors.

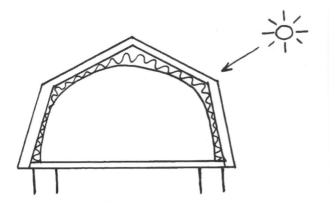

Mansard roofs offer both angles, and also minimum cooling surface and maximum habitable space: cast shade ratio.

Keypoints

- Drain-down solar water-heating systems must have their collector panels at least 60mm (2½") above the solar tank. For thermo-syphon systems, the reverse: tank at least 900mm (36") above the collectors.
- The taller the tank, the more efficient the system.
- Rectangular roof planes best suit rectangular solar water and photovoltaic panels.
- Optimum panel orientation is due south. But up to 45° deviation is acceptable.
- Optimum inclination is incident to equinox for solar water heating, around 15° shallower for PV – but isn't critical.

For water heating, flat-plate panels are cheaper but non-south-facing roofs justify vacuum tubes as these can be rotated to improve orientation. There are also self-contained thermo-syphon systems, which must be emptied in frost seasons, so are only for summer use. These systems are simpler, so less expensive than others, but offer only limited CO_2 reduction – and, most importantly, you mustn't forget to drain them!

Nowadays, both PV and solar hot-water systems are long-lived and maintenance-free (PV inverters only last some 11 years but, for the rest of the system, manufacturers claim 30-year lifespans). Many even have self-cleaning glass; although in dry weather, panels may need accumulated dust hosing off.

Even if neither system is affordable now, it's wise to allow for future installation. This affects where you put roof-lights and chimneys, and perhaps roof pitch. You also need to allow space for solar tanks, which are heavy – so usually best located on the ground floor. Most have a 600 x 600mm (24 x 24") footprint: but allow more for their enclosing cupboards. Tall tanks allow water to stratify according to temperature, so can accept lower temperature inputs on cloudy winter days; and outputs from their upper water layers can sometimes even be hot enough to heat skirting radiators or underfloor heating in winter. Consequently, the taller the tank, the more efficient is the system.

Choices

Options	Decision-making factors
Solar water heating, photovoltaics or both?	Roof area; boiler suitability for warm feed; budget; hot-water / electricity use-proportion; daytime electricity use.
Most efficient or cheapest panels?	Roof area; longevity (e.g. glass or acrylic).
All-year solar hot-water collection or summer only?	Night frost predictability; CO_2 reduction or bill reduction; cost; bother; all-year or summer location; access to drain.
Hipped roofs to minimise cast shade or rectangular roof planes to maximise solar panels?	Garden position and size; roof-space usability; priority: amenity, vegetable-growing or energy?

Wind energy

Small-scale wind energy certainly works. I've lived with it for several years, so know from experience – and also (now) know lots of things *not* to do! Although small-scale wind turbines produce less electricity per dollar than photovoltaics (the reverse of large-scale ones), they work at night and in cloudy winter weather, so the two systems are naturally complementary. Also you need more electric light in winter, when days are dark, nights long and there's more wind.

To work efficiently, wind generators need to be above turbulence. For this, turbine blades should be at least 10m (30') above the tallest obstruction (trees, house, etc.) within 100–150m (300–500').[11] This means tall towers. Although as expensive as the turbines themselves, towers greatly increase output and decrease bearing wear, so justify their cost. That's the theory:

it's why commercial wind-turbines stand so high. In practice, however, for domestic-scaled wind generators, tall towers bring many problems: cost, erection, visual impact, neighbour complaints, and planning permission.[12] This means compromise: towers should be high enough to avoid the worst turbulence; and designed (e.g. hinged at base) so that it's easy to de-mount turbines for maintenance.

In most urban situations, turbulence and wind speed reduction make small-scale wind-power a non-starter. Rooftops and chimneys may seem the obvious places to mount wind turbines, but this is inadvisable. Vibration can shake buildings – and, especially, chimneys – to pieces. And drive occupants mad! Tall towers (which, unless they are large in diameter, require long stays) don't suit constricted urban sites, but without them, there's so little wind and so much turbulence that urban wind turbines are typically only 5% effi-

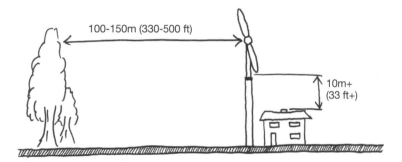

Wind turbines should be located clear of (or 30ft / 10m above) anything that causes wind turbulence. These are the minimum dimensions to ensure a turbulence-free zone. In practice, however, visual impact and expense may force a compromise. (Courtesy of Solarcity Inc.)

Turbulent zone: use a kite with fluorescent ribbons on its string to find the extent of turbulence. (Courtesy of Solarcity Inc.)

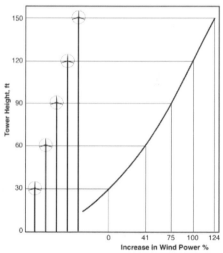

Height of tower and wind energy. The energy in wind increases with the cube of wind speed, meaning that a small additional investment in tower height can greatly increase energy production. (Courtesy of Solarcity Inc.)

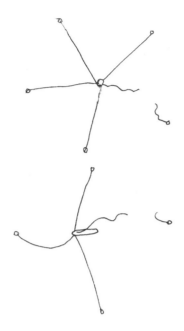

Five stays are much more secure than four. With five stays, if one breaks the tower still stands up. With four stays, it doesn't.

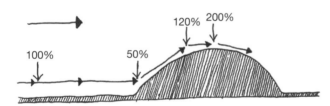

Wind speed at hilltops (Courtesy of Solarcity Inc.)

Caption: Turbulence at hilltops. (Courtesy of Solarcity Inc.)

cient.[13] They can even *drain* power, as inverters can draw off more than they produce.[14] Vertical-axis wind turbines suffer less than horizontal-axis ones from changing wind direction but, being less reliable in gusty winds, perform no better.[15] Moreover, some types rotate at unsettlingly frenetic speeds. For all these reasons, if you live in a town, wind-power may advertise your green aspirations – but don't expect it to deliver any benefits. For rural sites, however, it's worthwhile, and for off-grid locations, invaluable.

Low-price wind turbines produce low-voltage (12- or 24-volt) direct current (DC), so either need 12- or 24-volt DC systems (e.g. lighting made for cars, buses or boats) or inverters to change this to 110 or 230-50-volt alternating current (AC) for normal household appliances. Low-voltage cables need large – so expensive – conductor wires or they lose voltage over distance: so inverters should be as near generators as practical. Simple wind turbines have simple speed-governors, which can't be guaranteed not to break propellers in extreme gales, so, for safety, turbines shouldn't be too close to houses. Batteries, being 12- or 24-volt, should be on the generator side of the inverter, preferably in a detached shed due to acid-spill and fire risks. Batteries are expensive and have a limited lifespan, so are no substitute for mains connection (if available) as back-up and to feed back excess electricity.

Keypoints

- For wind-power to be effective, turbines must be above turbulence.
- Urban wind generators are rarely worthwhile unless towers are high.
- Expect to spend as much on the tower as on the turbine.
- Off-grid, wind generators are complementary to photovoltaics, and often invaluable.
- Batteries and inverters should be close to generators.

Choices

Options	Decision-making factors
Wind turbine or not?	Annual average wind speed; obstructions; tower height required; mains connection or off-grid.
Near house or on hilltop?	Distance; efficiency increase; cable cost; voltage-drop.

Hydropower

Hydropower uses flowing water to turn turbines. We're familiar with monster dams on major rivers – and the ecological damage they cause. But micro-hydropower uses small streams. Small systems only borrow a little water over a short distance, and overflow over knee-high dams is negotiable by fish: so they do much less ecological harm than large systems. I've lived with micro-hydropower for about 15 years. Hydropower is less weather-dependent than wind and sun as streams run at night: and most do all year. Consequently, its output – although perhaps varying seasonally – is constant. Additionally, turbines and alternators are easily accessible for maintenance – another great advantage over wind-power. Few people, however, have their own streams. If you haven't one, don't bother to read this.

Fairly small streams are adequate if there's enough height, as drop produces more energy than flow. (My own system has a stream about 1m (3') wide and 150mm/6" deep but, with 20m/65' fall, its Pelton turbine produces 1.3 to 1.5kw output.) Although most streams (like mine) for micro-hydropower are fast-flowing, larger-volume slow streams (which traditionally power watermills) can also be used.

Dams have two purposes: to raise water level above the intake pipe, or to impound water for when it's needed. Old waterwheels often have tallish dams to keep water throughout dry periods. Although larger mills worked all year, small on-farm mills usually only ground corn when weather was too bad (i.e. wet) for other work. For electricity, such intermittent production isn't worth it. Few micro-hydro systems, however, need big dams: just as well, as dam collapse can unleash a massive weight of water. Consequently, dams over 1.2m (4') high must be professionally engineered. Mine, however, is only 450mm (18") high: easy to build with boulders (using a 1.5m/5' bar to move the boulders around underwater, where their buoyancy in water makes them easier to lift).

The best way to prevent the pipe clogging with silt, autumn leaves or other debris is to use a Coanda screen. On the downstream face of a weir, this lets clean water through, while debris-laden overflow water continues down (or drains back to) the stream.

To prevent the turbine over-speeding and destroying itself, it needs either a constant electrical load or an output control device (e.g. adjustable sluice or inlet-restrictor). Sluices need to be automatically controlled from the turbine end, so necessitate unsightly overhead wires or expensive armoured cable in the piping trench. It's simpler, therefore, to maintain a constant

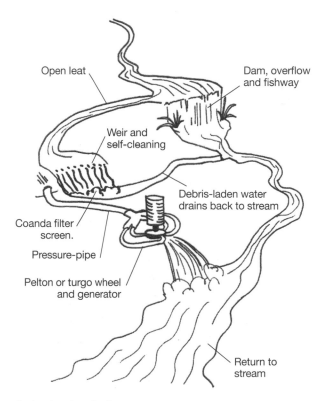

Open leat

Dam, overflow and fishway

Weir and self-cleaning

Debris-laden water drains back to stream

Coanda filter screen.

Pressure-pipe

Pelton or turgo wheel and generator

Return to stream

A simple micro-hydro system.

electrical load by selling excess power back to the mains (as long as there's a safeguard against mains distribution-system failure). Or shunt the excess into a ballast circuit (e.g. to heat oil-filled radiators or night-storage heaters). For safety, the ballast load (and fuses) must *exceed* the maximum the system could generate, at all times, without exception.

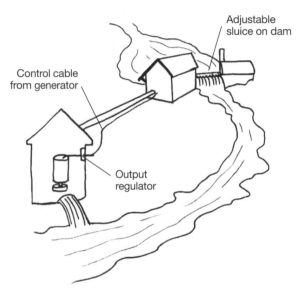

Over-speed protection by using an adjustable sluice.

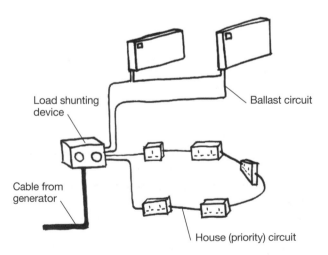

Over-speed protection by using a ballast circuit. (Any contact-breaker protection must ensure an electrical load remains.)

In my (1970s) prototype system, the pressure-pipe was as expensive as the Pelton turbine and generator together. Nowadays, mini-Peltons are much cheaper, but pressure-pipe is still expensive. It's cheapest, therefore to convey the water by a shallowly sloping open leat (clay-lined if necessary) or non-pressure drainage pipe, then a steep pressure-piped drop for the last bit.[16] Tempting as it is to reduce pipe size for economy, friction losses increase disproportionately with narrower pipes. It's best therefore to choose piping at least adequate for minimum flow; if you can add automatic inlet control to prevent air ingress, choose piping for maximum flow. (You need to avoid air bubbles because they effectively eliminate water pressure at the turbine. They can be carried down faster than they float up.) Additionally, self-cleaning leaf-screens (eg Coanda) help avoid air entrapment in autumn, when normal screens clog with fallen leaves. With these and lubrication-free bearings, weekly maintenance is eliminated, so the system runs mostly on its own: effortless free power.

Keypoints

- Hydropower runs all the time but output may vary seasonally.
- Turbines and alternators are accessible for maintenance.
- To prevent the turbine over-speeding, the system needs to have either an output-control device fitted (e.g. adjustable sluice, inlet-restrictor), or a constant electrical load (e.g. sell-back to the mains, ballast circuit).
- Drop produces more energy than flow.
- Dams over 1.2m/4' must be professionally engineered.
- Pressure-pipe is expensive, so minimise the length required by limiting steep drop to the last bit and only using pressure-pipe for this section.
- Pipe diameter disproportionately affects friction losses: choose piping at least adequate for minimum flow.
- Self-cleaning leaf-screens prevent pressure loss in autumn.

Choices

Options	Decision-making factors
Surface leat then steep drop or direct-route pressure pipe?	Land-form; slope; clay subsoil/availability
Pipe size?	Mains sell-back possibility; ballast-load heating need; low-flow adequacy; outlet control for low flow; air-ingress prevention efficacy
Ballast circuit, mains sell-back or inlet control?	

Heat pumps

Heat pumps are basically back-to-front refrigerators. Refrigerators cool things down and dump heat. Heat pumps cool something *else* down and *use* the extracted heat. They turn low-temperature heat (so low it would cool us) into high-temperature heat that can warm us: something for nothing – wonderful! However, this isn't as good as it sounds. They're electric-powered; and most electricity is fossil-fuel-generated. Moreover, it takes three units of primary fuel to deliver one of electricity: a ratio that must be taken into account when assessing heat pumps' usefulness. In the past, heat pumps' typical heat-output to power-input ratio (COP) used to be around 3:1. Although this sounded good, 3:1 x 1:3 made a net efficiency of 1:1, so they were in effect useless. In recent years, however, COP has improved markedly. Some Japanese ones now reach 4.9:1 COP.[17] The COP also depends on the output–input temperature differential. The lower this is, the less 'work' for the heat pump, so the greater its efficiency. Consequently, higher input-temperatures and/or lower output-temperatures can make heat pumps' installation worthwhile. Crucial to assessing their suitability, therefore, are input temperatures available and output temperatures required.

On the input side, the 'something else' from which heat is extracted can be ground, water or air. As heat sources, shallowly buried collector piping is sun-warmed and earth-stabilised. Ground temperatures approach the annual average temperature only 1–3 metres down.[18] Even at 1.5m (a common piping depth), the earth provides a huge reservoir of heat, which is replenished each summer. Shallow collector piping, however, cools ground all winter (when the heat pump is extracting heat) so prolongs ice on paving and retards plant growth in spring. This limits suitable locations. Solar heat can be extracted from under roads and car parks in summer and stored at up to 25°C (77°F) in the earth.[19] This is for groups of houses. Deeply buried collector piping is earth-warmed. Ground temperatures rise 3°C for every 100m (1.3°F/100') you go down. Deep boreholes aren't cheap, but can be shared between households.[20] Some areas (e.g. volcanic parts of New Zealand) can utilise geo-thermal heat near the surface. Piping below a compost heap (or, at a larger scale, in sewage) is biologically warmed. Heat pumps can also extract heat from water. Large ponds are reservoirs of low-temperature heat. Similarly, rivers' constant flow means that heat extraction barely cools their water. Many heat pumps extract heat from outdoor air. In damp, chilly weather, however, these can cool outdoor air to dew point and even freezing-point, so they ice up and don't work when most needed. To improve efficiency and counteract icing, some heat-pumps pre-warm air through solar panels (best mounted near-vertically on walls as winter sun is low). This, however, doesn't work at night, so your house needs sufficient overnight thermal-storage capacity.

On the output side, the temperature required from the heat pump depends on the room-heating method. Different heating methods require different temperatures. Whereas radiators need water at 60-80°C (140-176°F), radiant floors and walls only need water at 30°C (86°F)(to maintain a 19°C air temperature), and skirting radiators need water at 40°C (104°F). Heating a room with warm air, however, requires air (i.e. heat-pump output) at only a few degrees above room temperature – much lower than heating with hot-water systems. The more (i.e. the faster) the heat pump blows out warmed air, the closer to room temperature (i.e. the lower temperature) this output air needs to be. This decreased temperature differential increases efficiency, giving air-to-air heat pumps the highest COP. These, however, bring all the air quality problems associated with forced air heating (see: **Chapter**

13: What we breathe, page 131). Consequently, when COP figures are quoted, you need to know what input and, particularly, *output* circumstances they apply to.

Similarly, (genuinely) green electricity changes the arithmetic. Nonetheless, only where temperature differential is low and/or electricity is green, do I consider high-efficiency heat-pumps worth using.

Keypoints

- Heat pumps are basically reverse refrigerators, extracting heat from ground, water or air.
- With low output–input temperature differential, they can deliver energy savings.
- Air-source heat pumps can suffer from icing problems in damp cold weather.
- Ground temperatures approach the annual average only 1-3m (3-10') down and rise 3°C (6°F) per 100m (330') depth; more in volcanic areas.
- Shallow collector piping cools ground so prolongs ice and retards plant growth in spring: not good under gardens.
- Rivers, lakes and ponds are (low-temperature) heat reservoirs. Sewage water is warm.
- Warm-air output from heat pumps adversely affects air quality.

Choices

Options	Decision-making factors
Heat source: ground, water or air?	Ground temperature; use of ground surface; warmth of water; solar-warmed air; air humidity.
Output temperature?	Input temperature; renewable or fossil-fuelled electricity; solar-warmed air and overnight thermal-storage capacity.
Forced air heating, radiant floors and walls or skirting radiators?	Efficiency; health issues; comfort; price.

Combined heat and power

Combined heat and power (CHP, Co-generation) works in a similar way to car engines: heat-expanded gas powers a generator and the waste heat warms the home (or car). In cars, explosion gases turn an internal combustion engine, to which the generator is connected by a belt. In CHP systems, more slowly expanded gases power a Stirling engine, which turns the generator. As Stirling engines are externally heated, any form of heat, including solar, can power them.

CHP is really an industrial-scaled technology but home-scaled units are rapidly growing in popularity. In power stations, electricity production is the aim: heat is the by-product – normally wasted but usable for district heating. In district-heating systems, excess heat can be stored for several days and excess or inadequate power fed into or drawn from the electricity grid. In individual houses, however, electricity demand tends to come in spikes, whereas heat demand is more even.[21] Consequently, heating-need must determine output: electricity is the by-product, partly used and partly sold to your mains supplier. Although CHP seems appealing for off-grid homes, this is the wrong way round: if you have renewable fuels, heating is anyway easy but electricity production doesn't match demand profile. Indeed, CHP generators are designed to be run for long periods at a time, so heat slowly. For home heating, therefore, they should best charge an accumulator tank, from which heat can be drawn as and when needed. Slow heating means slow electricity production, so you must depend on battery storage.

Keypoints

- Combined heat and power systems can prioritise either heat or power, but not both.
- At a large scale, CHP units use waste heat from electricity production.
- At an individual house scale, CHP produces electricity from heat production.
- In individual homes, demand patterns combined with production method mean that heating-need, not electricity-need, should control output.

Choices

Options	Decision-making factors	
Individual-home or communal system?	Heat-storage possibilities	Allocation and payment; maintenance and management.
	Electricity storage	

Resources

General
http://www.energysavingtrust.org.uk/sites/default/files/reports/PoweringthenationreportCO332.pdf
www.icax.co.uk/thermalbank.html
www.otherpower.com/otherpower_wind_towers.html
www.solacity.com/siteselection.htm
www.solacity.com/SmallWindTruth.htm

Solar panel inclination
www.solarpaneltilt.com/#other

Less common products
www.aquashear.com/why-aquashear.cfm
www.solaredge.com
www.solaressence.co.uk/solar-pv

1. UK climate.
2. E.g. Solartwin: *www.solartwin.com*
3. There are also flexible PV membranes, but these aren't mass-market products.
4. The time of solar noon, however, varies some 25 minutes (hence about 7° in orientation) over the year: in Greenwich, 11.50 in December, 12.15 in February. On the summer solstice, it's about 4 minutes after clock noon, so approximately 1° west of south: *www.timeanddate.com/worldclock/astronomy.html?n=136* (accessed: 16 June 2014).
5. SAP2009 calculations cited by *www.solaressence.co.uk/solar-pv* (accessed: 29 March 2013).
6. Above 25°C (77°F), efficiency drops 0.5%/°C: Raferty, T. (2012) Green Roofs & Photovoltaics, Go Green, 2012, Bauder Ltd, Ipswich.
7. For some hours, not just at noon.
8. This is a very complex calculation, but the commonest rule-of-thumb is latitude minus 15°: *www.solarpaneltilt.com/#other* (accessed: 3 February 2015).
9. Modest shade reduces performance about 20%: SAP2009 calculations cited by *www.solaressence.co.uk/solar-pv* (accessed: 21 January 2015).
10. *www.solaredge.com* (accessed: 29 March 2013).
11. *www.solacity.com/siteselection.htm* (accessed: 11 December 2013); *http://www.otherpower.com/otherpower_wind_towers.html* (accessed: 16 May 2014).
12. With large-scale wind turbines, noise is an exaggerated concern (they're quieter than main roads). Domestic-scale turbines, however, are clearly audible close to. Far from being disturbing, though, their sound can feel reassuring.
13. University of Southampton Sustainable Energy Research Group, cited in *Specifier Review*, Issue 8/2009.
14. *www.solacity.com/SmallWindTruth.htm* (accessed: 11 December 2013).
15. Ibid.
16. Pressure-pipe takes pressure to join. You must lever sections together with a bar (and protective wooden block). If you have to solvent-weld any adaptations or repairs, don't do this in the trench. Solvent is heavier than air, so you'll breathe a whole lung-full, and, if you pass out, more lung-fulls till you stop breathing.
17. Due to Japanese energy-efficiency legislation: MacKay, D. (2009) *Sustainable Energy – without the hot air*, UIT, Cambridge.
18. 1m (3') in UK, 3m (10') in northern Sweden. By 6m (20') (in Britain) they're at a stable 11°C (52°F).
19. *www.icax.co.uk/thermalbank.html* (accessed: 17 May 2014).
20. *www.kensaheatpumps.com/products/shoebox-range* (accessed: 29 September 2014).
21. Ibid.

V. WHEN AND WHO ISSUES

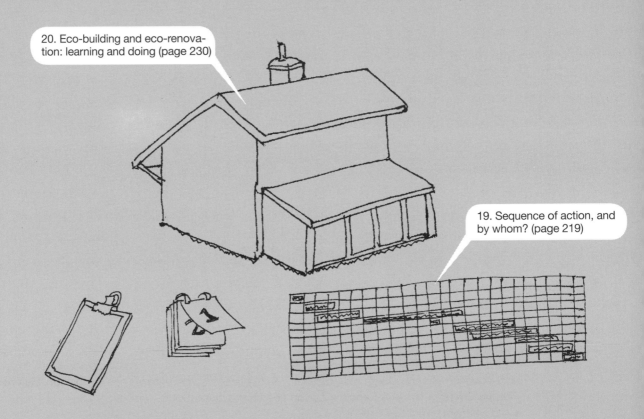

20. Eco-building and eco-renovation: learning and doing (page 230)

19. Sequence of action, and by whom? (page 219)

Sequence of action, and by whom?

CHAPTER NINETEEN

Where to live?
(page 224)

What order to
do things?
(page 224)

Who builds your
house?
You? (page 223)
With friends?(page 226)
A builder?(page 221)

Improving the
house while living
in it (page 228)

Procedural basics

Before building a new building, you need to get lots of permissions. That's obvious. But what about making changes to an old building? You may need some permissions for this too. Before doing anything, there are some things you need to be sure about:

- Is what you want to do legal?
- Is it safe to do?
- Will it compromise the building's safety and longevity?
- Will it compromise the building occupants' safety in a fire?
- Can you afford it?

Legality is mostly about planning permission, building regulations and safety issues. Although the foregoing general principles apply in many places, details, and the concerns that shape them, differ between countries, and sometimes states. As I can only describe the British situation, you'll need to check local planning rules and building codes before doing anything.

In Britain, planning is largely about appearance (e.g. planning guidelines, listed buildings) and use-changes (e.g. changing single dwellings into multi-apartments, shops into homes).[1] Some of this is open to interpretation, discretion or fitting into the right categories; some is also subject to a committee decision, so consent or refusal aren't foregone conclusions. Allow eight weeks for this process. If refused permission, you can reapply, changing those things that were listed as reasons for refusal. If you dispute these reasons, you can appeal – but keep this as a last resort, as the inspector's decision is final.

British building regulations are more firmly defined than planning rules, and conventional ways of compliance are explicitly described. (In contrast, US codes tend to be prescriptive; and Swedish ones, two-part: obligatory, and those required for loan approvals.) Among other issues, building regulations particularly cover construction durability, including structure, fire-spread and deterioration, and occupant safety. However onerous they may seem,[2] the regulations are clearly written, so you can (eventually) know what you're doing wrong and make the changes necessary for compliance. If you're using unconventional techniques (e.g. earth, straw), this process can be difficult and slow – but it's never impossible.

Other regulations (e.g. health and safety, electrical, gas) relate to the safety of both building workers and house occupants. Affordability includes financing arrangements, cash-flow and living expenses until you have a home to live in and time to do a money-earning job. Liveability is mostly about disruption (which depends on the nature of the building work) and your ability to cope with it (which includes both stress on family life and the scale and duration of the project). Buildability is both about the complexity and scale of the task and your abilities, energy (and perhaps strength) and endurance.

However, before thinking about the sequence of building work, you must decide *how* the building work will be done.

For both new build and renovation:
- Will you be doing the building work yourself?
- Will friends/volunteers be helping?
- Will a contractor be doing the work?

Or will you choose a combination of the above (e.g. perhaps you feel comfortable doing the less skilled jobs or less technical parts, or just labouring)?

Additionally, for a renovation:
- Will you be living in the house or somewhere else while the work is being done?

The answers to these questions raise many, very different, considerations.

If somebody else builds or improves your house on a fixed-price contract, your only concern is what you need to have done first. If the price *isn't* fixed or there are too many contingencies to allow for, however, you need to focus on what the builder *must* do (i.e. the things that *you* can't do but can't live without). If it's you who builds or improves the house, you need to think about where you'll store weather-sensitive or expensive materials and equipment (i.e. those worth stealing), somewhere to use as a workshop, and where you're going to live. If you'll be living in the house while the work is done, you need to think about what you need for a (semi)-normal life. If friends come from some distance to help, where will they stay, and do you have adequate facilities to feed them?

Before even planning any building work, however, you also need to be clear what your priorities are – as outlined in **Part II: WHY ISSUES**.

Keypoints

- Before starting work, consider legality, safety and affordability.
- Also, who will do it, where you can live and how you can survive the disruption and discomfort.
- Identify what your priorities are– and ensure these will be met.

Somebody else builds your house

This is the easiest way. Someone else works out what order to do things in – and does all the work. Not surprisingly, it's also the most expensive way. To control costs, it's wise to have a contract describing exactly what you want done, and preferably how, in terms of material, performance and finishes. The builder undertakes to do this for a fixed price. (Remember that trade prices are usually *before* sales tax (in Britain, VAT) is added. Contract prices are also usually pre-tax.) Once you, or your architect, have prepared the description (e.g. drawings, schedule of works and specification), you can get standard forms of contract (e.g., in Britain, JCT 2011 Minor Works Building Contract[3]). Contracts and specifications can be exactingly detailed and watertight (and expensive);[4] but for a simple, single house, fairly simple ones (e.g. peppered with phrases like "to relevant British Standards and Codes of Practice" and "workmanship to be to best practice standards") often suffice. The contract must, however, *fully describe* anything builders aren't familiar with. In English-speaking countries (Canada excepted) and those with brief or mild winters, this includes most energy conservation details. It's wise also to check that your builder is insured and unlikely to go bankrupt mid-job.

As building progresses, the builder may discover unanticipated problems (e.g. unrecorded underground drains, unavailable materials), and also you'll probably have fresh ideas – so you'll need both a contingencies budget (normally 10%) and an agreed formula by which to cost variations. The simplest formula is the builder's (or sub-contractor's) hourly rate added to the cost of material bought (or saved).

Small builders don't carry as many overheads or travel so far to the site as larger firms, so are usually cheaper. But builders and homeowners have different agendas. Homeowners want the work on their home done fast. Builders want to *stay* in work, so tend to have several jobs on the go at any one time – hence the not uncommon practice of doing a day's work, then disappearing but leaving sawhorses on the site to mark their ownership of the job.

Also, many small builders are loath to commit to a fixed price, so may only give estimates, and work on a time plus materials basis. This can be formalised as a 'cost plus' arrangement: actual costs (materials, labour, sub-contractors) and an agreed percentage for overheads, management and liabilities. Such arrangements should be cheaper than fixed-price contracts because they don't have to cover minor contingencies (which fixed prices usually do) – but they can be risky. Unfortunately, there are lots of cowboys, bodgers and downright crooks in the building trade, so it's essential that you *know* (from their satisfied clients) that your builder is trustworthy, competent and reliable. Besides the potential for dishonesty with 'time plus materials', if your builder departs from instructions, it could be *you* who pays for his time to rectify it. To protect against this, you'll need to have a contractual agreement (albeit with no fixed price) and means of ensuring the builder doesn't reimburse himself by charging more in other parts of the job. One way to do this is to break the job down into tiny work- and payment-stages, so it's easier to check the timesheets they submit. (Also, the more often they're paid, the easier is their cash-flow, so the sweeter they'll probably be.) Nonetheless, cost overrun is always a risk. To protect against this, you need frequent reviews – *and* to do the most essential things and anything you're not able to do yourself, first. To protect against things done badly, you can retain 5–10% for 6 months from completion, then only pay when all snags are cured. Estimates are exactly that: *estimates* – not fixed prices. So it's wise to add 20% to them: can you still afford this? Between estimates and fixed prices is the Target Pricing system. This limits risk for both householder/client and builder. It also, however, depends on trust and trustworthiness.

Whatever system you choose, when it comes to critical details (e.g. airtightness, cold-bridges), you need to discuss with your builder how, and in what order things should be done; and also decide at what stages things should be inspected and/or tested before any covering-up work starts. Airtightness is so uncommon in Britain and Ireland that it needs particular attention (see: **Chapter 7: Airtight construction**, page 70).

Target Pricing

Materials priced at cost.

Hourly rate for labour varied up or down – up to an agreed margin – should the job take less or more time than estimated.

Examples

Labour priced at (say) £15/hour, but varied up or down by up to 20% (here, £3/hour), should the job take shorter or longer than estimated.

Estimate: 100 hours @ £15/hour = target price: £1,500

Agreed margin: hourly rate plus up to £3/hour if job takes less time than estimated, or minus up to £3/hour if job takes more time than estimated.

Actual job:

A.   150 hours (50% more) @ £15 - £3 (maximum margin) = £12/hr
     150 x £12   = £1,800
     (you pay only 20% more: builder earns 20% less per hour)

B.   110 hours (10% more) = £13.50/hr (but absorbed within the 20% margin) so = £1,500
     (you pay target price: builder earns 10% less per hour)

C.   90 hours (10% less) = £16.50/hr (but absorbed within the 20% margin) so   = £1,500
     (you pay target price: builder earns 10% more per hour)

D.   75 (25% less) hours @ £15 + 3 (maximum margin) = £18/hr
     75 x £18   = £1,350
     (you pay only 10% less: builder earns 20% more per hour)

Keypoints

- A fixed-price contract best ensures cost control. Allow another 10% to cover contingencies. The contract must fully describe the work, especially anything builders aren't familiar with.

- Many small builders work on a time plus materials basis. If your builder is absolutely trustworthy and competent, this is usually cheaper, but cost overrun and departure from instructions are risks. Add 20% to the builder's estimate.

- Target pricing limits risk for both you and builder, but also relies on trustworthiness.

- In some countries (particularly the British Isles), most builders are unfamiliar with critical thermal-performance details, so these, and their inspection, need discussing in advance.

Choices

Options	Decision-making factors
Fixed price or target price contract or time plus materials?	Builder's trustworthiness; fixed-maximum or flexible budget.
What must the builder do? What can you do?	Standard required; your skills, time, energy, equipment and budget.

You build your house

Building takes time, energy and money. If you do it yourself, you're free to decide the balance between these. Usually, however, money is fixed and your energy is limited, so you have to put up with extended time. For self-building, you must have – or be able to pick up – basic skills, moderate strength, more than moderate endurance, preferably open-ended time and enough money to keep you over the time it takes. For skills and strength, average practical-mindedness and health normally suffices. I've led, trained and supervised volunteers of all physiques and all (anyway two) genders: some even with *no* previous skills. Although the difficulties shouldn't be underestimated, building or converting your own house isn't unduly difficult. (I've done both – and started-out with next to no knowledge of construction, building-process or site-management.) Moreover, self-building greatly reduces, perhaps eliminates, worry about being able to pay off your mortgage. The most important thing, however, is that you (as a family) *want* to do it.

If you decide to build or convert a house yourself, there are some basic safety rules:
- Never work so far to the side from a ladder that you must stretch or shift your body; and make sure the ladder can't slip (or tip when you're at the top).
- If working above the ground or handling heavy things, always have a clear space to jump to if things go wrong.
- Never lean brooms against low scaffold platforms or lorry floors. When someone jumps down, they can catch in trouser bottoms and slide up the trousers, causing a gruesome accident.
- When lifting something heavy, bend your knees, look up (to straighten the spine) and, in case you drop it, point your toes in (so it's deflected by your thighs).
- Wear steel toe-capped boots if handling heavy objects (e.g. concrete blocks, stones).
- Never leave nail-points sticking out of wood. Even if you'll be extracting them later, hammer them over.

- Cap (and perhaps flag) all sharp protrusions you could bump into: especially those at eye-level, as they're easy not to see.
- Keep anything dangerous to children (e.g. poisons, caustics, sharp tools, glass, etc.) securely beyond their reach.
- Always hold hand-held circular saws with both hands. The saw can jump at a knot and amputate the hand you're using to steady yourself. After that you're not steady, can't cut straight and may fall and hurt yourself.
- When big machinery is on-site (e.g. excavators, lorry delivery-cranes, concrete-lorry chute-booms), keep your distance.
- Protect your eyes when cutting blocks or painting overhead.
- And so on…

And one golden rule, never ever to break (which I have broken, to my lasting regret):
- Never work on a roof without somebody else (preferably with a phone) nearby.

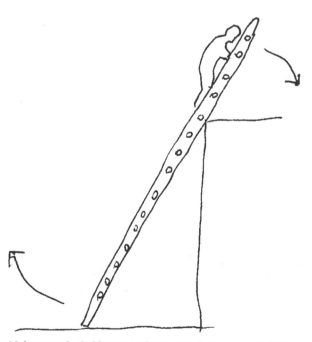

Make sure the ladder can't slip or tip when you're at the top.

What to build first, though? The first thing to *not* build is the house. First you need to build, buy or rent a shed (conventionally the garage) to store stuff in. It's also usable as a workshop and for making tea.

Next comes: where to live? On-site or, more comfortably, elsewhere? To live in, a caravan provides basic amenities, although it can be squalid in winter when it's full of wet clothing, so you might find it worthwhile to build a temporary open porch – yet another delay before you can start work. Once these practicalities have been worked out, only then are you ready to build. But, again, what should you do first?

The first thing of all is to get a water supply. No water means you can't mix concrete or mortar, nor wash tools. Even worse, you can't make tea – so can't be a proper builder! If there's electricity nearby, it's also very handy to get this connected. Electric cement mixers start easier than petrol ones – and they're quieter and don't give off clouds of smoke. Also, power tools make many jobs easier and quicker. Temporary water supply can be over-ground, but in frost season needs to be underground or protected (if temperatures barely fall below freezing, rags may suffice).

Unfortunately, enthusiasm doesn't guarantee efficiency. If your approach to work is overly led by enthusiasm, you start with things that show results and leave the invisible (drudgery) bits till later. Then winter sets in: short days, gloomy light, mud and slush, rain, wind, ice and snow. This dampens enthusiasm, making you start to wonder: perhaps it's time to quit? To avoid this, you need an efficient – rather than overly enthusiastic – approach.

For efficiency, you need to lay the groundwork first. This includes literal groundwork (e.g. planks to wheel barrows along); power and water supply; materials and tool storage; also site organisation, cleanliness and order. You also need space to work. It'll save much time and energy if you use mortar straight out of the wheelbarrow, bricks or blocks and timber straight off stacks, and stones from a flat-bed cart. Leave

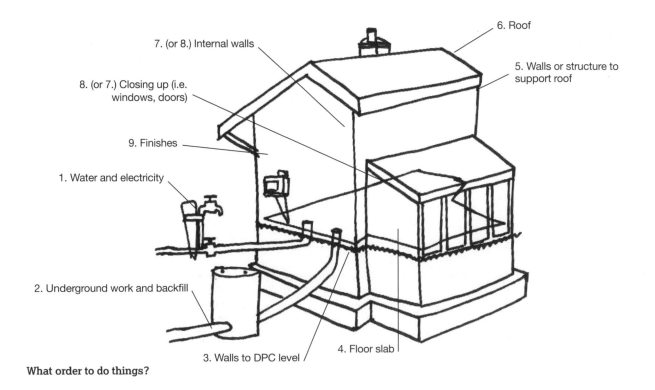

7. (or 8.) Internal walls

8. (or 7.) Closing up (i.e. windows, doors)

9. Finishes

1. Water and electricity

2. Underground work and backfill

3. Walls to DPC level

4. Floor slab

5. Walls or structure to support roof

6. Roof

What order to do things?

space for this. Also for scaffolding: the cheapest towers have 4 x 4' (1.2 x 1.2m) footprints; more expensive ones are larger – and sturdier. (Don't use the *very* cheapest: they're just for painting, so are too flimsy for building.) You'll need to be able to barrow past them, so add another metre of clear space.

Before you build anything above ground, you need to complete all underground work. Although it's demoralising to work for weeks and see nothing rising out of the ground, it's even worse to have to negotiate planks across open trenches and mounds of excavated earth – all soft, cloying, wet and slippery. Get the underground work (drains, water, conduits for future electricity, gas, telephone, etc, foundations, walls up to DPC-level and all backfill) finished, then pour the floor slab. You now have somewhere level, clear and clean to work off. You can then get materials (e.g. sand, gravel, bricks) tipped or delivered around it, in the right places to minimise carrying and wheelbarrowing. On sloping sites, get anything heavy put uphill, not downhill, of where it's needed.

All this preparation may be an (apparently) time-consuming, anticlimactic start – but it saves the time it takes many times over. By easing subsequent work, it lets enthusiasm grow as work progresses. I've started in both ways, so I know from bitter (and rain-soaked) experience.

Your next aim is to get the roof on. Once you have a roof, keep two 'work tracks' – indoor and outdoor – going at the same time (see: **You build your house with friends**, page 226). It may seem inefficient, but this way you'll never be held up by weather.

Then, sooner or later, comes closing up: hanging doors, fitting locks and glazing windows. If security and protection from weather are your main concerns, this is sooner. If the building needs to dry out, the later you do this, the better. (Even paint contains a lot of water, let alone concrete, mortar and plaster.) If you're still handling bulky or awkward objects (e.g. scaffolding) or even just lean brooms against windows, there's also a risk of breaking glass.

Keypoints

- Building or converting your own house greatly reduces mortgage worries.
- Building takes time, energy and/or money. Do you have a balance between them, or enough of one to cover lack in another?
- What is the cost to family life? Does your family support this route?
- *Always* follow basic safety rules, and also think of child safety.
- Don't build the house first but build, buy or rent a shed or garage to store stuff in and use as a workshop. Consider living in a caravan on-site.
- The first thing is to get a water supply, and preferably electricity. Then complete all underground work and backfill. Raise walls to damp-proof-course level so you can pour the floor slab: somewhere level, clear and clean to work off.
- Get materials delivered to near where they'll be used.

Choices

Options	Decision-making factors
What is the balance between time, energy and money?	Your time, energy and money.
Can you afford the time to build or must you work for an income?	Budget; your income-earning job; cash-flow; domestic disruption; site-attendance for deliveries.
When to close the building up?	Security; risk of breaking glass; weather protection; drying out; warmth to work in.

You build your house with friends

Building your house with others puts you in the position of manager, but as they're friends, you can't order them around. They're co-workers, effectively volunteers. Also, word will get around and people you've never met before will probably turn up and want to join in, as building is – at heart – a creative experience: not a drudge job. They want to experience this. But unavoidably there's plenty of drudgery too. So you'll have to arrange work so that they don't feel like unpaid serfs. In short, you need to manage the job in exactly the same way as for volunteer building. After all, if the experience isn't fulfilling, volunteers won't volunteer and friends won't stay friends for long.

There are three key foundations for fulfilment: motivation, satisfaction with the result and personal growth. Volunteers (and friends) must, therefore, feel their effort, hence the job, is worthwhile: so it needs to be 'deep-eco', not 'shallow-eco'. The end-result must be artistically satisfying: so must involve *their* creativity within *your* framework. And they also need to have experienced overcoming obstacles and learnt new skills: so you also have educational responsibilities. Secondary benefits include sociability – and related fun; and teamwork – and related camaraderie. This builds the experience that *meaningful work* can be *fun*.

Creative/artistic experience is vital but brings a whole set of problems: who makes decisions? Who will suffer the end-result? An imposed plan denies volunteers any creative input. A free-for-all lacks cohesion – and leaves you to live with it. Neither is satisfactory. Art, however, needn't be an individual-ego activity. It's possible to make *group* decisions (e.g. by mocking-up the positions and shapes of windows, or rehearsing the spatial and view experiences produced by different door-swings or turns of stairs). There's no difficulty in this; I've done it many times.

As making new friends is a major reward, every opportunity to do things together – from food-preparation to throwing, catching and stacking bricks on scaffolding[5] – is worth taking. Teamwork is one such opportunity. Many jobs have only certain bits that require skill, knowledge, strength or experience. Few amateurs have all of these, but teams often cover the whole range.[6] Moreover, work goes much faster, so everyone feels they've achieved something. Similarly, heavy, messy, unpleasant – and dangerous – work, especially in deluging rain and boot-sucking wet mud, is soul-crushing drudgery on your own. In a group, adversity builds camaraderie, and can even be fun.

Inexperience, however, means volunteers aren't used to anticipating risks. This necessitates extra vigilance and, specifically, ensuring that nobody works beneath anyone else or near big machinery. Mess is also potentially dangerous; besides hindering efficiency and being demoralising. Everyone has come to work, though, not just clean up. Cleaning up, therefore, must be a joint task, not one dumped on an individual. All this has implications for how you organise the site and manage work.

Neither volunteers nor friends are cost-free: they eat lots and work short hours. Nor should they be regarded as free labour: that's treating them little better than slaves. Nor do they lessen your hours of work. For teamwork to proceed without impediment, you may need to do all the fiddly preparatory work first: so no time to put your feet up after everyone else has stopped work. And, as nobody enjoys working in the rain or indoors in sunny weather, you must prepare *two* work tracks: indoor and outdoor. Using volunteers does save money and advance the project's completion[7] but to build a good spirit into your project, it's vital that material benefits for *you* are outweighed by fulfilment benefits for *them*.

Volunteers/friends must, of course, be organised: led. But this requires a style of leadership different from the norm. It must put worker-satisfaction to the fore, and, most importantly, you must never delegate unpleasant or dangerous work to others, but do it yourself – either with others or on your own.

Keypoints

- Don't treat friends/volunteers as free labour to exploit. Ensure they gain more than they give.
- For motivation, the project must feel worthwhile, not just be for your benefit.
- For their fulfilment, the work should offer opportunities for inner growth, teach new skills, and the end-product must be artistically satisfying.
- Maximise opportunities for socialisation and teamwork.
- Prepare indoor and outdoor tracks so workflow isn't interrupted by complicated fiddles.
- Be extra vigilant over safety matters.
- Never order someone to do a job. *Ask* them.
- Never, never, ask someone to do something you wouldn't do yourself. Indeed, if it's a really awful job, *do* it yourself.

Choices

Options	Decision-making factors
Smooth workflow or overcome challenges?	From which do they learn more and experience greater fulfilment?
Pre-empt problems or work out what to do on the spot?	Size of group; members' (and your) skills and experience.
Artistic free-for-all, coordinated, pre-planned or by consensual agreement?	Who will live with it? Learning through group-work; attitudes to bella-donnaism or consensus decision-making.

You buy an old house and someone else improves it

This is much the same as having somebody else build a new house, but there are additional issues: particularly unpredictable complications, costs and completion time.

Cost-wise, there are many more unpredictables with old buildings than with new-build, so builders tend to add more contingency money to fixed-price bids. If you're waiting to move in, how long can you wait? And what does the wait cost you? If you're already living in the house, how can you cope with the disruption? You can, of course, include late-completion penalties in the contract but, again, the builder will probably increase his fixed-price tender to cover for this. You'll also need to clearly define what causes of delay they are or are not liable for.

Construction-wise – and hence aesthetically – unpredictable faults and opportunities are also likely to show up. The older the property, the more you can expect. The builders can, of course, resolve (or obscure) the faults and plaster over the opportunities on their own. And, unless you have close contact with the job, they will.[8] This means wasted opportunities. Consequently, the more you're involved – but without getting in the way – the better should be the result.

Keypoints

- Builders tend to cover for unpredictable contingencies by inflating fixed-price bids.
- If you're living in the house, how can you cope with the disruption?
- What do construction delays cost you? The contract can include late-completion penalties, but this will probably increase fixed-price bids. For these, liabilities and exclusions must be clearly defined.

Choices

Options	Decision-making factors
Late-completion penalties in the contract?	Effect on price; your costs; clarity of applicability; acceptability to builder.
Penalty rate?	
Causes of delay that the contractor is or isn't liable for?	Definition of 'abnormal weather'; delays by others.

You improve an old house while living in it

This is the easiest to achieve: easiest site to buy, fewest planning permission problems, least expensive – and you have complete control over what you spend. However, it's also far the most exacting. What is the cost to family life? Will you end up with a house that saved half the improvement cost (although none of the existing house cost), which a divorce settlement forces you to halve; not to mention tearing you in half? If, on the other hand, both you and your partner support this route and both take part, this can be a bond-strengthening experience.

This way also puts the most demands on sequence. If you need to buy materials, when will you need them? Must they be special orders – so vulnerable to delays? Where will you put them? If you store sand and gravel in heaps, you need to cover these against leaves, twigs and cats – who treat them as a free lavatory. If you need to make new openings in walls,[9] or store steal-worthy materials outdoors, how will you keep everything safe? You also need to keep most things dry, so this is probably the first thing to attend to. (Even if they're undamaged by getting wet, they're heavier to handle and the house takes longer to dry out.) If you store dangerous (e.g. sharp, poisonous, caustic) or bad-to-breathe dusty things (e.g. cement, lime) indoors, how will you keep children safe?

Next comes making (but not necessarily finishing) the kitchen. This is the core of family life. In it you can make food, wash and, in extremis, sleep. It's probably the easiest room to keep warm. You'll need to keep one largish downstairs room (e.g. a future living room) as a workshop. Anyway, you're not likely to have time to watch TV: to survive as a family, it's more important to get the house straight as fast as possible. This unfortunately often means that lots of second-priority finishing-off jobs never get finished.

Keypoints

- Sequence is critical: both construction sequence and to free space to store things in, work in, live in.
- Plan when you need materials delivered, where you'll put them and what to do if special orders are delayed.
- Don't underestimate pressures on family life and child safety issues.

Choices

Options	Decision-making factors
Which part(s) will you live in? Which part(s) will you use as a workshop and materials store?	Children's safety; health risks (e.g. dust, mould, fibres, fumes); fire hazards; comfort; convenience; cleanliness; effects on family life.
What sequences of work and of completion should you aim for?	Efficiency; workspace; domestic space; disruption; allowance for let-downs by others.

Resources

General

Day, C. (1998) *A Haven for Childhood*, Starborn Books, Clunderwen, Wales.
Day, C. (1990) *Building with Heart*, Green Books: e-book from *http://gum.co/bftJ*)
Day, C. (2003) *Consensus Design*, Architectural Press, Oxford.
www.rics.org/uk/shop/JCT-2011-Minor-Works-Building-Contract-MW-18852.aspx

Mortgages for eco-projects:
www.ecology.co.uk

1. Even here, there are (mostly small) differences between the constituent nations, and between county planning policies. Nonetheless, the principle is uniform.

2. Since 1965, the building regulations are now about 10 times as bulky and cost some 100 times as much.

3. *www.rics.org/uk/shop/JCT-2011-Minor-Works-Building-Contract-MW-18852.aspx*

4. Architects and quantity surveyors will advise you that this saves money in the long run. This may well be true but there's also an element of "they would say that, wouldn't they?"

5. The technique is to scoop the brick at the apex of its flight and continue its momentum to the stack. And wear thick leather gloves.

6. Day, C. (1990) *Building with Heart*, Green Books: e-book from *http://gum.co/bftJ*)

7. Some say volunteer building costs more due to invisible costs (e.g. food, accommodation) but my experience is that it delivers buildings at one-fifth to two-thirds of contract cost. And imprints them with a priceless good spirit.

8. Notice I refer to builders as male. Most are. Many are strong-willed (as building is about overcoming difficulties); and some are macho. A very few all-women teams exist. These tend to be more sensitive to the homeowner's needs.

9. A warning: to demolish blockwork walls, don't use a sledgehammer unless you don't care about cracks throughout your house. Sledgehammers make short work but shake neighbours' houses too. A less destructive method is to chisel (or angle-grind or saw) out one block at the top of the wall, then use a club-hammer and bolster to prise off blocks one row a time.

CHAPTER TWENTY

Eco-building and eco-renovation: learning and doing

It's relatively easy to create a new eco-home. All it takes is forethought and under-standing. And, of course, money or – if you build it yourself – your energy. With old buildings, however, it's certainly harder (but definitely not impossible) to bring them up to the energy performance standards we expect today. In recognition of this, the PassivHaus Institute has lowered its Heat Demand condition for certifica-tion from ≤ 15 kWh/m².yr to ≤ 25 kWh/m².yr for EnerPHit (eco-renovation) certifi-cation. The PassivHaus standard represents a 90% energy-efficiency improvement over normal houses. Currently, reaching this costs around 20% more than to reach Code for Sustainable homes: Level 4.[1] It doesn't have to, though: a few have been built at comparable cost, and Jon Broome has designed a £45,000 self-build PassivHaus.[2] But what if even that is too expensive? With new buildings, an 80% improvement in energy saving is a reasonably affordable – so accessibly cost-effective – target. Most of this is achievable with older buildings – albeit at the price of disruption. The first 40–60% is relatively easy. However, even a 10% improvement is worthwhile. With lower aspirations, though, it's important to avoid, or at least minimise, 'lock-in' (see: **Chapter 11: Insulating old buildings: basics**, page 115). In short, meeting current eco-standards with old buildings is certainly harder than with new-build. Harder, but by no means impossible.

For both new and old buildings, there is, of course, much, much more to eco-design than any single book can cover. One essential point is that energy conservation shouldn't be confused with ecological design. Energy isn't the same as ecology. As a habitable climate and life-supporting oceans depend on energy conservation, it's an urgent priority, so a central part of eco-design – but *it's only a part*. Ecology, after all, is about life, not just energy.

I hope, therefore, that I've made it clear both that eco-design involves many more issues than energy; and that it's easy to do, whatever your financial resources and level of skill. Despite the subject's complexity, the recommendations in this book are enough to achieve very significant improvements in all the most important aspects. As every situation is different, however, there can be no simplistic formula or set of prescriptive instructions. Formulae applied in the wrong circumstance are useless – sometimes worse than useless. Understanding is a much sounder foundation for action. I've tried, therefore, to make the issues simple to understand – so you know what to do in whatever circumstances you find yourself.

From many years of experience, I've learnt that lots of things can go wrong. I haven't glossed over these. Indeed, you can't expect to create anything worthwhile without encountering problems and making

mistakes. So, if you want to find reasons why things *can't* be done, there'll always be plenty to choose from. The corollary, however, is that once you decide to do something, problems tend to evaporate. Indeed, if you want to create an eco-home, it's easily achievable even with (as I had when I started) limited money, knowledge, skill and endurance to overcome obstacles.

In this light, I advise you to look at the mistakes others, including myself, have made and learn from them (see: **Case studies: Eco-home examples,** and the **What I should have done** reviews, page 232). Foremost among these are mismatches between design and climate. To adapt the Norwegian saying: there's no such thing as bad weather; only bad design.[3] Observation of what weather does, how this affects us, and how nature and design traditions respond to this helps avoid 'bad design'. Adding an understanding of *how design affects us* can raise our designs from the merely 'not bad' to health-giving 'good' design – healing to both people and planet.

I've also learnt that eco-design is not just affordable, but actually saves money in the long run. More importantly, it feels personally fulfilling – and creates a home to which you feel much more connected than had you just bought an 'ordinary' one.

And for the planet and all who live on her, it's essential. Now you are ready to start. Good luck.

1.　De Selincout, K. (2014) How to save social housing blocks, PassiveHouse+, 2014, Issue 8.

2.　The materials cost £40,320; the rest is specialist labour: Broome, J. (2014) The £45,000 low-energy home, *Green Building* magazine, Autumn 2014, Vol. 24, No. 3.

3.　'There's no such thing as bad weather; only bad clothing.'

Case studies

Eco-home examples

These are examples of eco-homes I've designed, and in some cases built with my own hands – so, from direct experience, I know what is wrong with them. They include, however, two projects that haven't yet been built. Why do I show these? One exemplifies cooling in a cruelly hot climate; one taught me many social lessons; and one (still unfinished) demonstrates my current thinking. All these projects are listed chronologically, so you can see how my thinking progressed, and what I learnt from each.

Ruin (ex-chapel) conversion in west Wales, 1971

This house is extremely exposed to mountain weather, especially wind. By 1970s standards, it's highly insulated, but poorly by today's. It has wind-turbine electricity (unknown in the UK at that time). As the original building was very ruined (below knee-height in places), the whole interior and rear extension is new. It has a turf roof (also unknown outside Scandinavia at that time, so improvised). The main (and bad weather) entry is to the sheltered north. It has four compact bedrooms (two downstairs, two in the roof): all opening off the living room. It has a fireplace at one end and a Rayburn cooking stove in the sunken kitchen (600mm/2' lower) at the other. The extension contains a bathroom, larder and study.

Front view.

View from the garden.

Cross-section showing entrance sheltered by hedgerow and rising land to the north side.

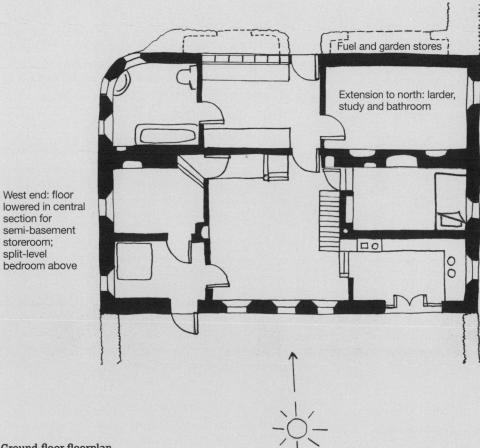

Fuel and garden stores

Extension to north: larder, study and bathroom

West end: floor lowered in central section for semi-basement storeroom; split-level bedroom above

East end: main floor lowered for kitchen and bedroom; bedroom above

Ground-floor floorplan.

Review: What I should have done

The combination of compact design and two central chimneys (each warming four rooms) with the Rayburn at lowest level (so heat rises from it to the rooms at a higher level) keeps the house snug. The Rayburn was the best choice at that time (not as efficient as an Aga, but smaller and easy to light, so not necessary to keep alight all year), but it was designed to burn coal (which I now – but didn't then – know is the most CO_2-producing fuel); there are now even more efficient *wood*-burning alternatives (no net CO_2-production). Also, the open fireplace should have had a glass door (which also weren't available then, but are now). The worst heat loss isn't (as anticipated) from the north or east, but by ferocious (although mild) southerly gales. A conservatory (with PV roof) along the whole south façade would overcome this, but would be visually unacceptable in this wild landscape. Instead, the large south windows should have had internal shutters (which could open into the deep reveals). Even with the large south windows, the 4.6 x 4.6m (15 x 15') living room is too deep for adequate daylight from those windows

alone: it needs a light-tube at the roof ridge. For disabled accessibility (not considered then), the entrance, living room, bathroom/toilet, study and one bedroom are on one floor, which is good. But the kitchen is three steps (600mm/2') down, making it inaccessible without a ramp, and the bathroom should have been a wetroom.

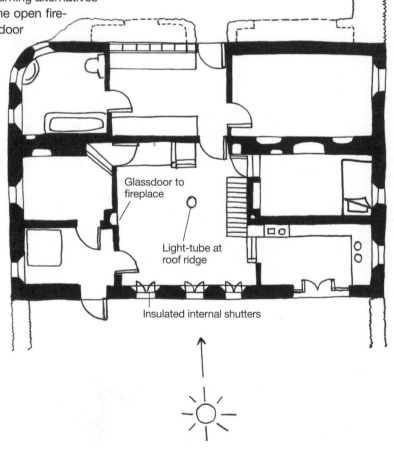

Glassdoor to fireplace

Light-tube at roof ridge

Insulated internal shutters

Improvements to the design.

Farmhouse conversion in west Wales, 1976

This farmhouse is exposed to wind and mountain weather, but sheltered to south and west by being in a slight dip; to east by (new) windbreak trees; to north by an (existing, but heightened) hedgerow. The house is very compact and organised around a (new) central chimney for a Rayburn cooking stove. One upstairs bedroom has a sleeping loft. The dining/kitchen opens on to a stone-walled sun-trap.

As this room is along the south wall, the larder is internal (see illustration on page 161). The existing stone walls were in poor condition, so externally insulated, then stonework-faced. There are lift-off internal shutters to windows, and hydro powered electricity and solar hot water: both improvised prototypes. My office, in the former hayloft, was warmed by cow heat from below.

House and office (over cowshed).

Central chimney for Rayburn cooking stove.

Cross-section.

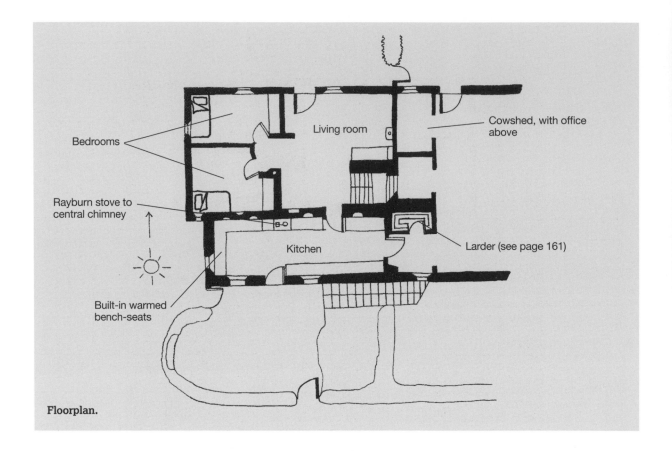

Bedrooms

Living room

Cowshed, with office above

Rayburn stove to central chimney

Larder (see page 161)

Kitchen

Built-in warmed bench-seats

Floorplan.

Review: What I should have done

Visitors describe this as the warmest house in the region. It reduced energy consumption by about 75%. But its stove mostly burnt coal (fortunately extremely little), rarely wood. (The two other stoves were wood burners, but never needed to be lit.) Severe rain-exposure means the south-facing soft-wood doors repeatedly swell and shrink, compromising draught-seals. For disabled accessibility, the step-free entrance, living room and two bedrooms are on one floor, but the short (1.2m/4') passage to bedrooms is too narrow for a wheelchair, and these and the kitchen are one step down (to minimise the building's height in the landscape).

The bathroom/toilet, one bedroom and study are upstairs. To adapt for disabled access, therefore, I had to convert my workshop into an apartment with a wetroom. Being outside the original thermal perimeter, so not heated by the Rayburn, this depends on (new) central heating. After the cow died, my office, also being outside the thermal perimeter, needed (and got) an insulated floor and woodstove. Outdoors, the sun-trap yard has a gate in the corner (to eliminate contact between damp hedge-bank and house), which denies a protective corner for breakfast-time sun. The gate should have been differently placed.

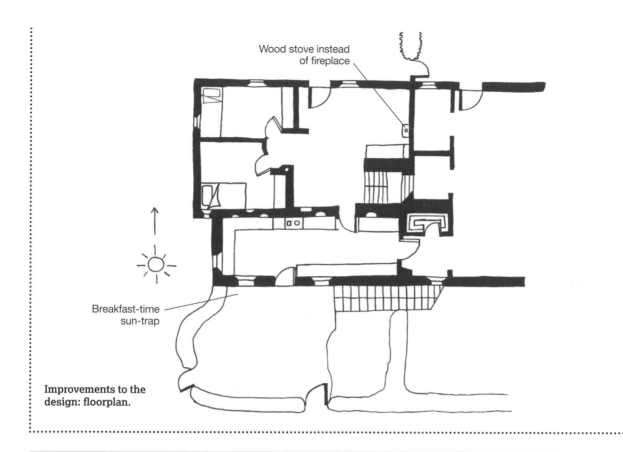

Wood stove instead
of fireplace

Breakfast-time
sun-trap

Improvements to the
design: floorplan.

Farmhouse conversion in west Wales, 1978

This house is located in a sheltered valley bottom; the land around it is a bit damp. It's organised around a (new) central chimney for an Aga cooking stove. For solar warmth, there's a conservatory all along the front wall, with solar water-heating panels (improvised from radiators) all along its roof. Under this are roll-down insulated blinds.

The conservatory for solar heating doubles as both greenhouse and entry.

Review: What I should have done

As this house's stone walls look attractive internally and are part of local character externally, they aren't insulated. (Although stone is a poor insulator, traditional stone walls are rubble-filled, so insulating airspaces make them warmer than 1980s houses.) Also, double-glazing wasn't then available. In theory, the glazed-roof conservatory would make the house too hot in summer: but as it's shaded by tomato-plant foliage, this is rare. Aga stoves take long to light (due to massive thermal capacity), so tend to be kept burning all year: this constant warmth was never unwelcome in 1978, but by 2020 will be. A long ex-cowshed is adjacent to the conservatory; with a bank/hedge on the other side of the house and control of the gates, this could flush the house with cool night air in summer. In the 1970s the need for cooling was inconceivable, so the gates and doors aren't optimally positioned. As the surroundings are damp (the house's name means 'boggy pool'), this house would have benefited from clay plaster internally.

Extension to a row of farm buildings in south Wales, 1988

This house is in a bleak ex-mining area, windswept in winter but sheltered by a steep slope to the north-east. It has a fully glazed sloping south façade, with kitchen, bedrooms and wetroom to the rear. It too is organised around a (new) central chimney for a Rayburn cooking stove.

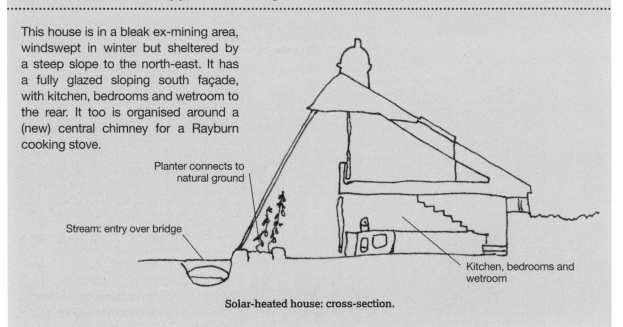

Planter connects to natural ground

Stream: entry over bridge

Kitchen, bedrooms and wetroom

Solar-heated house: cross-section.

Review: What I should have done

I last visited in 2003, and heard praise for its warmth and no reports of overheating, but our climate has noticeably changed since then. It should have a generous (controllable) ridge vent and roll-down hardwood-slat shades externally and insulated fabric shades internally. Also, well-insulated shutters would be better than curtains to divide off the conservatory from the other rooms. Were the streamlet shallower, so broader, it would reflect more winter sunlight indoors. The natural-earth-connected planter is a potential source of moisture in the air, but doesn't appear to have caused any problems.

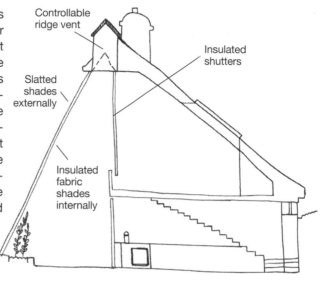

Controllable ridge vent

Slatted shades externally

Insulated shutters

Insulated fabric shades internally

Broader stream to reflect sunlight

Improvements to the design to allow for cooling in increasingly hot summers.

Shed conversion, west Wales, 1990

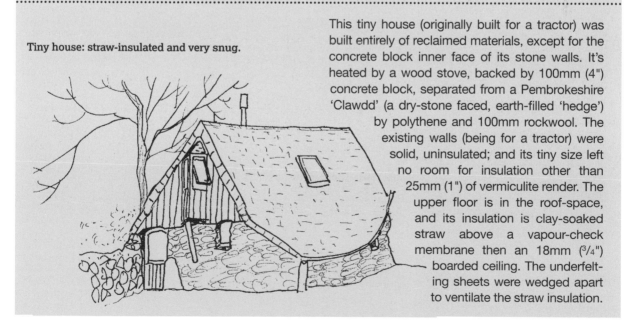

Tiny house: straw-insulated and very snug.

This tiny house (originally built for a tractor) was built entirely of reclaimed materials, except for the concrete block inner face of its stone walls. It's heated by a wood stove, backed by 100mm (4") concrete block, separated from a Pembrokeshire 'Clawdd' (a dry-stone faced, earth-filled 'hedge') by polythene and 100mm rockwool. The existing walls (being for a tractor) were solid, uninsulated; and its tiny size left no room for insulation other than 25mm (1") of vermiculite render. The upper floor is in the roof-space, and its insulation is clay-soaked straw above a vapour-check membrane then an 18mm (3/4") boarded ceiling. The underfelting sheets were wedged apart to ventilate the straw insulation.

Review: What I should have done

As the tractor entrance was from the north, this was the only place for a sizeable ground-floor window: hardly ideal! Despite this and minimal wall insulation, this house is very snug. The thermal mass of the wall behind the stove keeps heat overnight; and the concave opposite wall, although only poorly insulated, reflects radiant heat back on to occupants. The roof insulation, however, is noisy. Diligent mice keep searching for the few remaining seed-heads in the straw. I should have dusted the straw with lime, which neither mice nor I like to breathe,[1] but I hadn't anticipated the mouse problem.

Eco-village, northern Sweden, 1992

The layout of this eco-village was designed jointly with the aspiring residents. Together, we modelled it in clay, each person locating and shaping their own house.[2] (There are also a communal meeting room, workshop and vegetable-store and private storage-compartments.) We then used an angle-poise lamp to check for sunlight throughout the year, and adjusted the form accordingly. The horse-shoe layout admits spring sunlight (in winter there's only brief twilight), maximises informal child supervision, security and wind-protection and is socially focusing. The community lawn gets evening sun-light in summer (from the north-north-west). Although (kakelugn) woodstoves provide the main heating, all homes have generous south glazing for solar heating in spring and autumn with 150mm thick insulated internal shutters for the dark winter months. All homes have disabled-accessible ground floors and expandability and divisibility potential to cope with future needs. It was intended that I would design each individual home jointly with the family concerned. In the event, we only designed one before the 1992 recession scuppered the project.

Climate-responsive design: buildings shield against wind but admit spring sun.

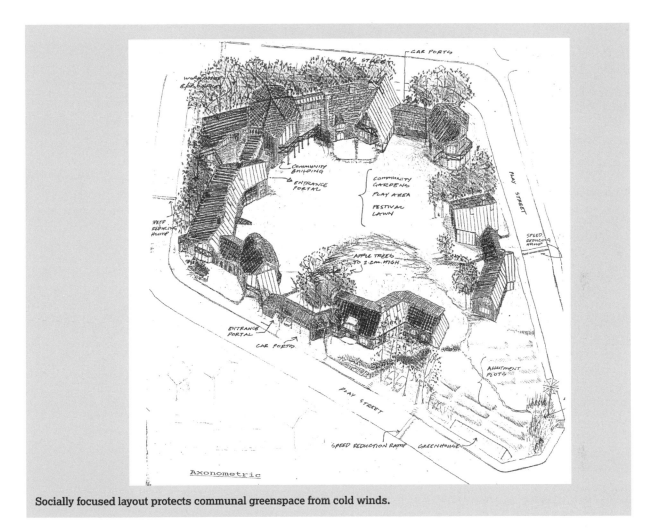

Socially focused layout protects communal greenspace from cold winds.

Review: *What I should have done*

As this was never built, I can only guess at its short-comings. The stream crossing the road was definitely a mistake: although channelled in narrow slots, abnormal weather would have spilled water on to the road, which would have remained as ice all winter. Some of the passageways into the central space were narrow and, although they were bush-shielded to minimise wind, they would have increased the carrying distance for snow shovelling as there would have been no space to heap snow at the side. Also, due to the low sun angle, the sloping glazing would have been no advantage: vertical glazing would have made windows and doors easier to seal. (This, however, would only affect the conservatories, not the house interiors, as insulated shutters give another layer of protection.)

Apartment over office in California, 1994

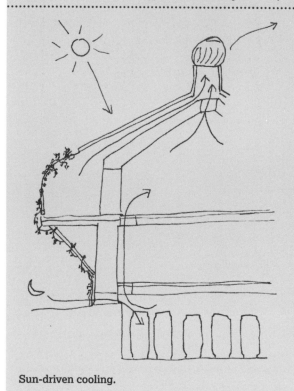

Sun-driven cooling.

Unlike all the others, this building is designed for natural *cooling*. It's in central California, where summers are exceedingly hot (routinely 114°F/45.5°C), and winters frost-free. Its construction is straw-bale, and water heating is solar. Insulation obviates the need for winter heating. For cooling, dark-tiled roofs taper to vent chimneys, so hot airflow (above the insulated ceiling) drags out room air. Replacement air is drawn from the basement, which is filled with water drums to store night-time coolth. The balcony is vine shaded. In a city with high crime statistics, windows and balcony give a view of all entry points and risk-locations on the property.

Review: What I should have done

Construction postponed, so no feedback. I'm fairly sure the *principles* work, but less sure about the *numbers.* For instance, would the warm-air extract and cool-air replacement rates have been sufficient? Should the roof-void and extract opening have been larger, and the solar chimney higher? As, in that area at that time, there was no heating/ventilating engineer interested in natural cooling, my sizing was based on guesswork. After all, you can't *see* the effect of friction on air movement: it's one of those very few things where calculations are more reliable than experience.

House in Leicestershire village, 1996

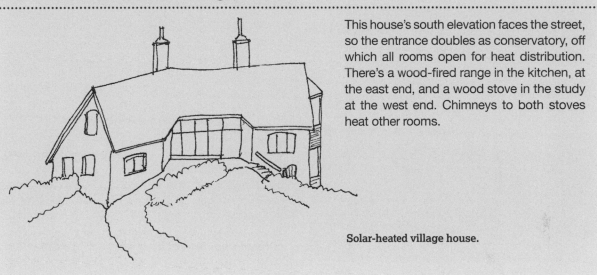

This house's south elevation faces the street, so the entrance doubles as conservatory, off which all rooms open for heat distribution. There's a wood-fired range in the kitchen, at the east end, and a wood stove in the study at the west end. Chimneys to both stoves heat other rooms.

Solar-heated village house.

Review: What I should have done

The owners have been happy with this house (but their neighbour didn't want *any* house there, and God forbid a wind turbine anywhere nearby!). I, however, have my doubts about the design, because of the conservatory at the centre. It's good for heat distribution and to buffer fluctuating temperatures, but to go from one room to another through it at night in winter could be chilly. There were, of course, insulated shutters – but these depend on conscientious operation: neglect to close them and you have a cool room as the circulation hub.

Large house in north Wales, 1998

This house has breathtaking views to the east, so glazing for solar heating and for views is quite separate. To minimise visual both impact on the land-scape and wind cooling the plan is spread out. Thick insulation compensates for the increased cooling surfaces.

Photographs: courtesy of Peter Saunders

Review: What I should have done

The building has a long perimeter so that every room enjoys stunning views, even though this increases cooling surface. Increasing the wall insulation from 200 to 300mm (8-12") compensated for this. The upstairs rooms are all fitted into the roof-space, which reduces the cooling surface and helps to shed wind. This meant that draught-sealing the vapour-check and wind-barrier membranes required great care. The changes of floor level focus the eye on different views and add delight, but for disabled accessibility would require several stair-lifts. Although there's space for this, a platform lift connecting all levels would have been better. I should, therefore, have included a storeroom on each floor, which could have the intervening floor cut out, for conversion into a lift-shaft. The conservatory was too hot or too cold some times of the year, so should have had (and now has) an opaque, fully-insulated roof.

I quote the owner's appraisal, with which I agree:

"The house is delightful with form and function in perfect tune. Indeed, it's a very harmonious place in which to live. We did feel that the conservatory was too hot in the summer months and too cool in the winder months, but found the solution in replacing the glass roof with a slate roof. We named the house Cynefin which is a Welsh word implying 'sense of place' which feels entirely appropriate for such a sensitive design."

**Glass conservatory roof replaced with slate.
(Photograph: courtesy of Peter Saunders)**

Apartment and Goethean science laboratory in Scotland, 2000

This building sits just below a hilltop at 200m (650') altitude, so suffers strong and bitter North Sea winds from the north, although existing larch trees give partial shelter. It is of timber construction with sheepswool insulation, clay plaster, a larch shake roof† and limecrete foundations. All wood and stone (for the base walls) is local. The solar-heated glasshouse is lower than the main laboratory so hot air can rise into the whole building. Between the glasshouse and laboratory are vertical-sliding sash windows and insulated shutters on the outside of a heat store (water drums, buried in cob). Besides this solar heating, it has a wood stove. An apartment, with kitchen below and bedroom in the roof-space, is at the eastern end.

Goethean science laboratory with apartment at east end. (Photographs: courtesy of Life Science Trust)

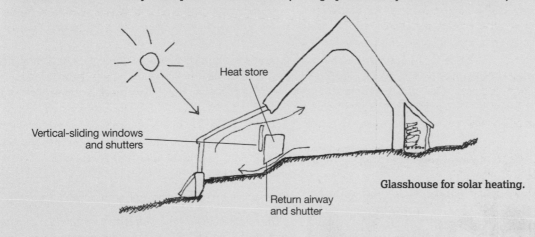

Heat store

Vertical-sliding windows and shutters

Return airway and shutter

Glasshouse for solar heating.

Review: What I should have done

Only recently completed and occupied, so I haven't yet had feedback. The several levels limits disabled access. But as the track to the building does so anyway, the situation is no worse.

New village house in Poland, 2012

(Also see illustrations on pages 138, 147)

This house is in central Poland. The climate is continental with 30°C (86°F) summers and sometimes -30°C winters; little wind, often none at all. For solar heating, it has south-facing sliding-folding glazed doors to the living room, backed by sliding insulated shutters. There's a wood stove with an under-building fresh-air inlet; also underfloor heating from a gas boiler, which also heats a 300mm (12")structural spine wall (consisting of the same type of piping used for underfloor heating with 100mm (4") brick leaves on either side). The central chimney contains both stove flue and all passive vent ducts. On the roof are PV and solar hot water panels. The ground floor is fully disabled-accessible; and upstairs bedrooms have sleeping-lofts.

House for hot summers, cold winters.

Review: What I should have done

Not built as site purchase fell through. I now think the south-facing roof-light would admit too much heat in summer, so it should have been a dormer.

Solar-heating conservatory extension to semi-detached house in Wales, 2014

This extension is on an extremely constricted site and its height is also acutely limited. The location is urban/suburban and near a hilltop, so there's ever-present wind; otherwise, the climate is generally mild. For thermal storage, it has exposed reclaimed brick walls (with 100mm/4" rockwool in their cavity) and a brick-pavoir floor over concrete and 100mm (4") insulation. For cold bridge elimination, all windows, door and (cavity-fill) wall insulation are in one layer, with a steel corner-post structure outside this. Against overheating in summer, there's reflective insulation in the roof; and, for cooling, door and opening window positions are carefully chosen to obtain through-draught from breezes diverted by surrounding buildings. This conservatory is intended to provide heat and pre-warmed inlet air to the whole house in winter and, in summer, to act as a solar-warmed chimney to cool the house at night.

Conservatory for solar heating.

Review: What I should have done

Not yet completed so too early for thermal evaluation. Shade from neighbouring buildings means no solar heating at winter solstice – when skies are anyway usually grey. Thereafter it will get good sun, as it faces, and the roof rises towards, due south. I'm very cautious of roof valleys, as they're prone to clogging and leaking. But, as I'm even cautious of more flat roofs, headroom here leaves no alternative, and I have given close attention to roof details. Space and roof drainage limited the insulation to a (high embodied energy) PIR warm roof. The brick walls and tiled floor are good for storing (and stabilizing) heat but neither are local: local works have been bought-up and closed-down by larger companies, and reclaimed bricks weren't available. As these (and so much glass) could sound hard, there will be soft furnishings in the shady areas (where they don't obstruct thermal storage). For summer cooling, the roof overhang blocks most sun; if required, short bamboo blinds hung from the soffit would block all till late afternoon. Whatever the wind direction, the carefully the positioned windows and door flush the room. I accepted the fact that the sash window, although draught-proofed, can never be airtight, but chose it as it doesn't obstruct the path past the building. Also it gives more versatile ventilation and direct-sun penetration options. I have not yet had a winter to test whether this was the right choice.

1. I describe this process more fully in *Consensus Design*, Architectural Press, Oxford.
2. I could, of course, have worn a face-mask, goggles and rubber gloves – luxuries denied to mice – but also encumbrances to work.

Glossary of technical terms

Glossary entries are indicated by[†] the first time they occur in the book.

Accumulator tanks: large water-tank heat-stores.

Batt: panel of flexible insulation, usually 450mm (18") wide.

Biodiversity: a broad range of living organisms. There is concern that nature's full range be preserved. Currently, it isn't: species extinction is accelerating.

Breather membranes: these stop liquid water penetrating the construction, but let water *vapour* evaporate from it. They're commonly used instead of (vapour-impermeable) bituminous-felt under tile or slate roofs, and as wind-barriers in wood-frame walls.

Brown roofs: roofs surfaced with rubble and/or loose stony subsoil, so attractive to insects (and the very few colonising plants that can survive on the non-water-holding soil).

Caravan: a wheeled home with towbar; in America, called 'trailer home'.

Carbon-negative buildings: buildings which export more energy (produced by renewable technologies, e.g. photovoltaics) than they import, so lessen CO_2 production.

Cold bridges: localised heat-loss routes, where insulation is absent or significantly inferior to that around it.

Coolth: the opposite to heat: a desirable quality in hot climates/weather.

Decibel: a unit of sound on a logarithmic scale, measured by air pressure. When measuring environmental noise, decibels are usually 'A-weighted' to reflect annoyance levels (the unit for this is 'dBA'). Noise is a subjective assessment, so not precisely quantifiable. 6 decibels represents a doubling of air pressure, hence (effectively) of noise.

Degrees Kelvin (K): A measure of temperature, equal in magnitude to °C but starting at a different zero point. In the context of thermal conductivity/resistivity, degrees K indicates a temperature difference. Also written K, Δt or δt.

DPC: damp-proof course, (nowadays) usually plastic or bituminous felt, both used horizontally and vertically.

DPM: damp-proof membrane, usually polythene (recycled is available) or (sometimes) bituminous paint.

Dry-lining: plasterboard fixed to walls (usually with dabs of plaster). In American: dry-wall.

Electric field: a field between an electric potential (voltage) and a negative one or earth, proportional to the difference in voltage. It's easily shielded by common building materials.

Electromagnetic field (EMF): a combination of electric and magnetic fields, produced by current *flowing* in a conductor, proportional to the load (i.e. voltage x amperage). It's a serious health concern (although disputed), but impossible to (economically) shield – so protection from it needs distance.

Eutectic salts: salts that dissolve into, or condense from, liquid at certain temperatures, so store heat or coolth in latent form. Because this heat is stored at close to surrounding temperatures, little is lost.

Flashings: pieces of folded metal (usually lead) that cover junctions between roofs and chimneys, abutting walls, etc.

Gabions: galvanised steel mesh cages for filling with earth, stones or rubble: usually used as retaining walls.

Geo-pathic radiation: natural terrestrial radiation distorted by underground water channels and rock fractures. It's another serious health concern (also disputed).

Green roofs (also called 'living roofs'): surfaced with growing plants, from sedum to flowers and meadow grasses – even bushes and trees.

g-value: the amount of solar heat admitted through glazing.

Habitable rooms: rooms used for living activities (e.g. living rooms, bedrooms); not kitchens, toilets or passages.

Hardboard: a pressed sheet of wood-fibres, bonded solely with lignin from the wood; 1,200 x 2,400mm (4 x 8ft) usually 3mm ($^1/_8$in) thick. 'Masonite' in American.

Heat-exchangers: these extract heat from outgoing air to warm incoming ventilation air. The two airstreams remain completely separate to avoid pollution transfer.

Heat pump: effectively a refrigerator run back-to-front. It extracts large-volume, low-temperature heat from ground, water or air to produce a smaller amount of higher temperature heat.

Home zone: a pedestrian-priority street (or group of streets) designed or adapted to meet the needs of residents, particularly children. Traffic is discouraged (although not forbidden) and vehicle speed restricted. Developed from the Dutch 'Woonerf'. Around the world they are known by other terms such as living streets, complete streets, shared zone, residential zone, etc.

Hygroscopicity: the ability of a material to attract and hold water molecules within it. Consequently, hygroscopic materials help regulate humidity.

Inner (and outer) leaf: the 'leaves' of a cavity wall.

Intumescent paint: paint that swells into an air-excluding (hence fireproof) layer when heated.

Joists: structural timbers supporting floors and/or ceilings, to which these are fixed; usually 50mm (2") wide, and of different depths according to span.

Kacheloen (in Swedish, kakelugn): a thermally massive stove that stores, then radiates, heat all day from one brief, very hot burn.

K-value: (also called 'lambda value') thermal transmittance, measured in watts/m.°K, (°K, here, meaning temperature difference in °C).

MVHR: mechanical heat-recovery ventilation to recover outgoing heat through heat-exchangers.

Mesoclimate: the climate of an intermediate-sized geographical area (e.g. city, river valley).

Microclimate: the climate of a small geographical area (e.g. garden, street).

Multifoil insulation: two or more sheets of reflector-foil laminated over air- or gas-bubbles or flexible insulation (usually fibrous) to provide multiple heat- (and coolth-) reflective surfaces.

OSB (oriented-strand-board): 2,400 x 1,200mm (8 x 4') varying thickness sheets of pressed wood fibres and glue.

Passive vent: a ventilation extract driven solely by natural air pressure. The outlet pipe is usually 125mm (5") diameter and must extend at least 1.5m (5') above the ceiling of the room it ventilates. Humidistat-controlled versions are available.

PassivHaus (plural: PassivHäuser): a building (of any type) conforming to the PassivHaus Standard – namely that only needs 15kWh/m²/year to heat, and is so airtight that unintended air-changes don't exceed 0.6/hour. To achieve this, PassivHäuser usually rely on MHVR.

Phase-change material: as eutectic salts, but not necessarily salt. Styrene wax is commonly used.

Photovoltaics (PV): the conversion of sunlight into electricity, usually by roof-mounted panels.

Phyto-toxic(ity): poisonous to plants.

Plasterboard: plaster-cored, paper-faced sheets, 2,400 x 1,200mm (8 x 4'), usually 9.5 or 12.5mm thick (other thicknesses and sizes are available, but less common). In American, called 'sheetrock', 'gyproc' or 'gyprock'.

Plenum: a space, usually above a ceiling, for pressurising air.

Purlins: beams running horizontally to support rafters; usually 75mm (3") wide, with depth according to span.

Rafters: structural timbers running up and down roofs; usually 50mm (2") wide, with depth from 75mm (3") upwards, according to span.

Reveals: the sides of window and door openings.

Roof-eave: the bottom edge of a roof, the overhang under it and the fascia board.

R-value: thermal resistance/metre thickness, measured in m². K/watt or (in USA) in imperial units; (K, here, meaning temperature difference in °C in UK, °F in USA).

Services: in American, 'utilities'.

Shake roofing: hand-split wooden shingles.

Soffit: the underside, particularly of eaves and window/door openings.

Softboard: board made of compressed wood-fibres, without any glue, usually 10–12mm thick.

Space heating: heating indoor space.

Stack effect: the upward movement of air due to vertical pressure differences resulting from thermal buoyancy. Warm air inside the building, being less dense than cooler outdoor air, seeks to escape from openings high up in the building, and is replaced by cooler (hence denser) air entering through lower-level openings. This process is driven by the combination of temperature difference and stack height. Chimneys are everyday examples.

Stud: vertical structural timbers, usually 100 x 50mm (4 x 2").

Terraced houses: in American: 'row houses'.

Terrazzo: concrete ground smooth to expose aggregate pattern. (Power-floated concrete can be equally smooth and hard, but is just a uniform grey.)

Thermal bypass: external air bypassing the insulation (so negating its effect) but not necessarily entering the house interior.

Thermal mass: the ability to store a substantial amount of heat or coolth without undue rise or fall in temperature.

Trimmer: a structural member that runs perpendicular (and usually in the same plane as) joists or rafters to support ones cut off at an opening (e.g. staircase, chimney, roof-light).

U-value: An element's heat-loss rate, measured in watts/m².°K/hour, (°K, here, meaning temperature difference in °C).

Vapour-check: a membrane impermeable to air, positioned to prevent moisture-laden indoor air reaching the insulation (where the moisture can condense into water). (It's called 'vapour-check', not 'vapour-barrier' in recognition of the fact that construction may partially compromise its impermeability.)

Vapour pressure: the amount of vapour (e.g. water-vapour) in air.

VOCs: volatile organic compounds. These are mostly solvents (hence volatile) and, being organic, are easily absorbed by, and react with, our bodily chemicals.

White glass: (also called opti-glass) glass with low iron content. This gives superior transparency.

Zero-carbon buildings: buildings which balance their CO_2 account by exporting as much energy as they import.

Index

Also by Green Books

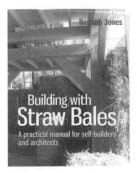

Building with Straw Bales: A practical manual for self-builders and architects

This book explains in straightforward terms the principles of straw-bale building for self-builders, architects and construction industry professionals. Written with non-experts in mind, this practical book takes you through everything you need to know in an easy, accessible way.

This revised and expanded third edition brings the book up to date and includes lots of stunning full-colour photographs throughout to illustrate the design and build process.

Also by Christopher Day:

Dying: or Learning to Live (Trafford Publishing, 2010)
ISBN 978-1-4269-2359-3

Environment and Children (Architectural Press, 2007)
ISBN 978-0-7506-8344-9

Places of the Soul (Routledge, 3rd ed. 2014)
ISBN 978-0-415-70243-0

Consensus Design (Architectural Press, 2003)
ISBN 0-7506-5605-0

Spirit & Place (Architectural Press, 2002)
ISBN 978-0-7506-5359-6

A Haven for Childhood (Starborn Books, 1998)

Building with Heart (Green Books, 1990)
e-book from http://gum.co/bftJ

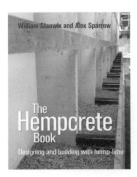

The Hempcrete Book: Designing and building with hemp-lime

The Hempcrete Book is a detailed practical manual for architects, surveyors, professional builders and self-builders. It explains how to source and mix hempcrete and how to use it in new builds and restoration. In colour throughout, fully illustrated with beautiful photographs, this book provides a full explanation of construction techniques, highlighting potential pitfalls and how to avoid them. It includes a comprehensive resources section and examples of completed builds, with design notes.

The Hempcrete Book is a valuable tool for any eco-builder.

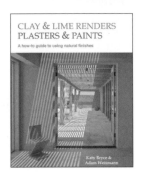

Clay and lime renders, plasters and paints: A how-to guide to using natural finishes

Clay and Lime Renders, Plasters and Paints is an in-depth guide to the selection, mixing and application of lime and clay based plasters, renders, paints and washes.
- Step-by-step instructions for applying lime and clay based plasters, renders and paints.
- Information on the benefits of natural finishes for personal health, the environment, and for buildings.
- Drawing on traditional methods & materials for using lime & clay finishes on new and historic buildings.
- A comprehensive and up-to-date online resource guide to suppliers, practitioners and courses.

About Green Books

Environmental publishers for 25 years.

For our full range of titles and to order direct from our website, see **www.greenbooks.co.uk**

Join our mailing list for new titles, special offers, reviews, author appearances and events:
www.greenbooks.co.uk/subscribe

For bulk orders (50+ copies) we offer discount terms. Contact **sales@greenbooks.co.uk** for details.

Send us a book proposal on eco-building, science, gardening, etc.: see
www.greenbooks.co.uk/for-authors

 @ Green_Books   /GreenBooks